Computer Modeling
of
Complex
Biological Systems

Editor

S. Sitharama Iyengar, Ph.D.
Associate Professor
Department of Computer Science
Louisiana State University
Baton Rouge, Louisiana

CRC Press, Inc.
Boca Raton, Florida

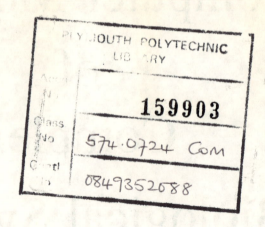
Library of Congress Cataloging in Publication Data
Main entry under title:

Computer modeling of complex biological systems.

Bibliography: p.
Includes index.
1. Biological systems—Mathematical models.
2. Biological systems—Data processing. I. Iyengar,
S. Sitharama.
QH323.5.C645   1984   574'.07'24   83-2686
ISBN 0-8493-5208-8

Direct all inquiries to CRC Press, Inc., 2000 Corporate Blvd., N.W., Boca Raton, Florida, 33431.

© 1984 by CRC Press, Inc.

International Standard Book Number 0-8493-5208-8

Library of Congress Card Number 83-2686
Printed in the United States

PREFACE

Computer technology has an important role in structuring a biological system. The availability of modern powerful computers has increased the development of good and accurate models of biological systems. Simulation may be carried out either with a digital or analog computer. Hybrid computers, which are a combination of both analog and digital computers, are used to simulate the behavior of the biological system.

Biological systems are subjected to physiological complexities and it is often difficult to predict behavior during experimental investigation. Furthermore, biosystems do not behave or act as expected. However, computer models can be helpful in understanding the behavior of complex systems. Over the past 10 years many researchers have been working on the development of models for biological systems. Examples of complexities like disease and cancer growth are very difficult to model in a system unless they are studied very well by physiologists or biologists.

The first chapter discusses the importance and applications of computers in simulation and modeling of complex biological systems with an overview of how computers may be used in ecology, physiology, drug receptor interaction, biomechanical and neurophysiological systems, thermal regulation in living bodies, and cell analysis and hematology. The second chapter considers the characteristics of biological systems and software and hardware considerations with an example of computer simulation of hydrodynamic shearing of a DNA biosystem. Chapters 4, 5, 6, and 7 consist of computer modeling examples of complex biological systems such as cardiovascular systems, cancer systems, drug systems, and food process engineering systems. The application of statistical techniques in modeling of complex systems with special reference to single and multiresponse models is discussed in Chapter 3.

The editor has made every effort to obtain contributions from the best qualified authorities from many nations; both he and the publisher would like to express their gratitude for their cooperation.

This book is written primarily for the benefit of graduate students and professionals who are interested in an interdisciplinary or multidisciplinary approach to complex problems of biological systems. Others who will find it valuable are biologists, chemists, engineers, research physicians, and computer scientists.

Finally, the editor wishes to acknowledge the assistance and encouragement of Dr. Walter G. Rudd, Professor and Chairman, Department of Computer Science at the Louisiana State University, Baton Rouge, Louisiana. Every paper in this book consists entirely of the author's opinions which may or may not agree with those of the editor.

S. Sitharama Iyengar

THE EDITOR

S. Sitharama Iyengar, Ph.D., is an Associate Professor in the Department of Computer Science at the Louisiana State University, Baton Rouge, Louisiana. His research and teaching interests in computer science include design of efficient algorithms, data-structure techniques, knowledge-based information systems, modeling and simulation of systems. Dr. Iyengar has published over 40 papers in national and international journals and chaired sessions in numerous international conferences.

CONTRIBUTORS

Nguyen Phong Chau, Ph.D.
Professor
Unite d'Enseignement et de Recherche
Saint-Quentin, France

W. Düchting, Ph.D.
Professor
Department of Electrical Engineering
University of Siegen
Siegen, West Germany

E. V. Krishna Murthy, Ph.D.
Department of Applied Maths and School
 of Automation
Indian Institute of Science
Bangladore, India

Pedro Francisco Castillo Monteza, Ph.D.
Department of Food Science
Louisiana State University
Baton Rouge, Louisiana

Stephen Quave, M.S.
Department of Computer Science
Utica College
Utica, Mississippi

M.S. Rao, Ph.D.
Visiting Professor
Department of Chemical Engineering
Louisiana State University Agricultural
 Center
Louisiana State University
Baton Rouge, Lousiana

Ramu M. Rao, Ph.D.
Department of Food Science
Louisiana State University Agricultural
 Center
Louisiana State University
Baton Rouge, Louisiana

Michel E. Safar, Ph.D.
Unite d'Enseignement et de Recherche
Saint-Quentin, France

M. Sahin, Ph.D.
Department of Biology
Semmelweis University of Medicine
Budapest, Hungary

TABLE OF CONTENTS

Chapter 1

COMPUTERS IN SIMULATION AND MODELING OF COMPLEX BIOLOGICAL SYSTEMS

M. Sahin, S. Sitharama Iyengar, and Ramu M. Rao

TABLE OF CONTENTS

I. INTRODUCTION

Simulation and modeling are important investigative techniques in any research activity for they provide a methodology for the design, development, experimentation, analysis, and evaluation of an experiment under study. Simulation which in plain words mean feigning of a particular situation provides a common base of techniques for the study of a diversity of projects and is a means for investigating a large number of ill-defined but solvable problems. Simulation provides a challenging atmosphere which is conducive to problem solving although it approximates the actual structure of the system. Simulation may be carried out either with a digital or an analog computer. Analog computers are mostly used for the study of behavior of mechanical systems and as such their use is limited to problems related to mechanical and/or electrical engineering. Digital computers are electronic data processing devices with incredible speed. Consequently, they have wide applications in biology, chemistry, agriculture, sociology, and in other branches of natural sciences and humanities. Hybrid computers which are a combination of both analog and digital could be used to simulate more complex problems. Computer systems, whether analog, digital, or hybrid, because of their high degree of modularity and collective complexity are natural objects of simulation models. For example, a month of computer processing can be simulated depending upon the complexity of the model on one computer run of less than 30 min.

The heart of simulation is the precise mathematical description of the system to be simulated. There are many reasons why a mathematical expression is preferred to a verbal explanation to describe a system. One reason is that several systems involve many different processes occurring simultaneously. The natural language is limiting when used for mathematical description of the system. Many of the most important behavioral aspects of complex systems such as nonlinearity, redundancy, and hysteresis cannot be explained in verbal terms, whereas a mathematical description of these concepts is often both compact and precise; this can be illustrated with a few examples.

Suppose we are interested in building a mathematical model which can predict the population growth of a particular town or city. The town or city in this instance can be considered as a biological system and let "N" be the number of people at a given time, "t". Then the rate of growth of the population with respect to time, is dN/dt. The rate can be calculated in a different way. Let us assume that there are no exogenous variables such as war, famine, or pestilence affecting the system. Under these assumptions, the rates of increase by birth and decrease by mortality are constant. Obviously the number of births is proportional to the number of people alive, similarly the number of deaths. If the birth rate is "b" and the death rate is "d", the total number of births is b.N and deaths is d.N. Therefore, in a time δt, we have (bN.δt) and (dN.δt) births and deaths respectively.

The change in population = δN = (bNδt − dN.δt)

$$\frac{\delta N}{\delta t} = (bN - dN) = (b - d)N = \alpha N$$

dN/dt = αN, where α is the excess of the birth rate over death rate.

A mathematical model similar to the one above can be constructed for any biological system where natural growth or decay occurs. Consider the production of certain antibiotics produced from bacterial cultures which are grown in controlled conditions. The growth of bacteria or microorganisms is governed by the same natural formula:

$$(dN/dt) = \alpha N$$
$$dN/N = \alpha dt$$
$$\log N = \alpha t + c$$
$$N = e^{\alpha t + c} = e^c = Ae^{\alpha t}$$

To evaluate the constant A, we use some unknown initial conditions: Say N = N, when t = 0

$$N_0 = A e^0 = A$$
$$N = N_0 e^{\alpha t}$$

This is an exponential law (+ve index), and so the rate is ever increasing. In the real world situations, these assumptions may not hold.

A second example to demonstrate the application of mathematical modeling in biological systems is the growth of bacterial cells in a culture medium. Suppose, the number of bacteria in a culture medium grew at a rate proportional to the number present initially. If in an experiment, it was observed that in 1 hr the number grew from an initial 100 to 332, and we are required to build a mathematical model to predict the number of bacterial cells at the end of $1^1/_2$ hr, then, let at time "t", the number of bacteria be "x"; dx/dt = rate of bacterial growth; this rate is proportional to the number of bacteria present, namely "x" is positive. So, we have,

dx/dt = kx; $x = A e^{kt}$ (1)
Initially it was 100;   (t = 0, x = 100)   (2)
It grew to be 332, in 1 hr;   (t = 1, x = 332)   (3)
To find the number, when t = 1.5   (4)
$100 = A e^0$; A = 100; $x = 100 e^{kt}$ (5)
$322 = A e^{k.t} = 100 e^k$; $e^k = 3.32$ (6)
  When t = 1.5, $x = A e^{(1.5)k} = 100 e^{1.5k} = 100 (3.32)^{3/2}$
  $\log_{10} x = 10g\ 100 + (3/2)\log (3.32) = 2 + (3/2)(0.5211)$
  $= 2 + 0.7816 = 2.7816$
  x = 60.47 = 605

In the preceding paragraphs, we have often used the word "system". An understanding of the meaning and characteristics of a system is approriate before we talk about "system modeling".

The term system is used in various ways by various people to cover many uses under differing contexts. However, in broad terms, a system may be defined as that arbitrarily chosen portion of space which is under discussion. All else is called the surroundings. The system is separated from the surroundings by either an imaginary or a real boundary line. To avoid uncertainty regarding what is specified as the system, its boundaries must be defined precisely. This decision may depend upon the purpose of the study. Closed systems are those across whose boundaries matter does not pass; in open systems matter does pass. The state of a system is fixed by describing the properties of the system at any given time. A process is said to occur in a system when any sort of change or transformation takes place. These changes are due to certain interactions taking place among various entities within the system. An entity in a system is defined as an object of interest which has separate existence. The property of an entity is called its activity. Therefore, the state of a system at any given moment can be defined as the description of all entities, attributes, and activities of that given system. Activity in any process is that which causes change in the system. During the process, the system changes from an initial state to a final state through a series of intermediate states. These series of intermediate states are called the paths of the process. Anything that crosses the boundary line and enters into the system is called the "input" and the ones that leave the system are called the "output". When the activity that causes change within the system can be described in terms of its input, the activity is called deterministic whereas stochastic activities are those which vary randomly and the output is not characterized by the attributes of the input. If the rate of change in a system over a period of time is constant, it is called a continuous system and if changes are discontinuous, they are called "discrete systems". Free systems are either wholly continuous or discrete. A homogeneous system is one in which the attributes of the entities are the same throughout and in a heterogeneous system the attributes of the entities vary from point to point. Endogeneous variables account for the activities of entities within the system and exogeneous activities are from entities outside of the system. Open systems have exogeneous activities and closed systems have endogeneous activities.

A. System Modeling

Modeling may be defined as the construction of a prototype, either mathematical or verbal, which approximately describes the behavior of a system under study. Behavioral models are especially needed where experimentation is physically or economically not practical, such as a situation study which has not been completely defined or a plan which is yet on a drawing board. There are two approaches to the study of behavior: one is experimental performed in a laboratory and the other is modeling. The models should be as abstract as possible and still be predictive. In other words, a model is something that mimics closely and foretells the relevant features of a system under consideration. The performance of a model is measured by the accuracy with which the model can predict the characteristics when applied to the system for which it was designed to handle. Accurate prediction by the model is affected to different degrees by numerous factors such as (1) proper design which includes proper specification of the system, surroundings, and the boundary lines and (2) identification and definition of individual and interrelationships of endogeneous and exogeneous entities some of which may be dependent variables and other independent variables. Absence of any information which is essential for construction of the model will lead to certain assumptions being made. The conclusions drawn from such a model will be greatly affected by the kind of assumptions made and also the input which enters into the system. A false or an unrealistic assumption leads to wrong or invalid conclusions. In brief, it can be stated that the main function of a model is to predict the performance criteria of a system under a set of conditions. Other benefits of modeling include: the design of meaningful laboratory experiments, evaluation of conflicting experimental results, and possibly offering an explanation for the differing results. If properly executed, a model should be much more efficient than either a theoretical or experimental process taken alone.

B. Model Types

Models are mainly classified into two groups: (1) physical and (2) descriptive. Descriptive models may be expressed in native languages or in terms of mathematical symbols to describe the status of variables in the system and the way the variables change and interact. Mathematical modeling needs a good knowledge of calculus.

Physical models are based on physical properties or comparison between mechanical, physical, or electrical systems and may be floor plans of a home or an industrial complex, pilot model plants of a distillation column, or instruments or means to measure the mechanical properties of a material. The advantage of a physical model is that it can be explained to any individual with limited technical knowledge. However, physical models are expensive to build, have limited use in that the model can be used only for the particular problem for which it was designed, and offer very narrow and unimaginative information to the decision-making process. Verbal descriptive models have limited communication and sometimes cannot be replicated. However, these models are the least expensive and so have found common uses in the decision process. Mathematical models mimic the conditions of a system in mathematical language or in precise mathematical formulas, which are concise and can be manipulated with ease. Besides, mathematical modeling can use numerous theorems which are available and can use high speed computers for quick calculations. Theorems are useful in drawing conclusions from simple models and computers are useful in drawing specific conclusions from complicated models. Furthermore, mathematical analysis of a system facilitates the construction of a tentative hierarchy, whereby each of the dependent and independent variables are rated according to the degree of their activity on the system.

Here, we are exclusively concerned with only mathematical models based on digital computers of certain complex biological systems.

C. General Methods in Building a Model

Model building is as much an art as it is a science. It involves intuition, imagination, and skill. It is impossible to state a set of rules to build a mathematical model as much as it is not possible to draw a picture or paint a landscape following a list of regulations. It depends upon the viewpoint and judgment of the modeler to decide which information should be included or emphasized and to what extent in the model. However, it is possible to offer a set of guidelines or a framework around which the modeler can develop and improve his skill and imagination to build a model. Besides, experience and common sense are the other ingredients of a good model process. The guidelines or framework are very general and applicable to most systems. The specific characteristics of each system determines its own framework which should be explored by the modeler himself. The following are some of the considerations to be remembered prior to any modeling process:

1. Understand the system and its components of which the model is to be built. These include the system structure, the system entities and their activities on the system undergoing a change or transformation, interrelationships among various entities, dependent and independent variables, system boundary and its surroundings, exogenous and endogenous parameters, etc.
2. Define in clearly understandable language the objectives of the model, what it is supposed to accomplish, what data are given, and what additional data or information is needed.
3. Review your model building methodology more than once and obtain an answer as to the method under consideration will accomplish the objectives for which the model is to be designed.
4. Make a thorough literature review on system models to determine whether any modeling or approaches suggested by other investigators for systems which are closely similar to having analogous characteristics of the system under consideration. A good literature review process should not only benefit the modeler in having a better grasp and understanding of the system under his consideration, but also will forewarn him of some of the obstacles, bottlenecks or surprises which he may encounter.
5. Classify and formulate the given data. Wherever possible, reduce the verbal data into mathematical or statistical symbols or formulas. Sort out the relevant data, facts, information, or logical parts of the problem from the superfluous.
6. Examine what mathematical or statistical theory or law is applicable to process the given data and to arrive at a solution.
7. Identify what additional data are needed and how and where to go about getting these data to complete the model.
8. If the modeling process involves a large complex system, try to individualize parts of the problems which may finally lead to one comprehensive answer. In order to do this, divide the large system into smaller blocks and represent the interrelationships of the sub-blocks by arrows. Such a pictorial block-arrow diagram of a complex and large system will facilitate easy understanding of the complexity of the system.
9. Check each sub-block or part of the system for its characteristics, its significance in the overall problem, its logical position in relation to other sub-blocks, its individual contribution in the final solution, and to what extent the individual sub-blocks affect the quality of the final model designed.

10. Synthesize results of each part of the model to verify whether the final model corroborates the individual solution.
11. Pay particular attention to the fact that the model proposed is sufficiently flexible to accommodate varying input/output data.
12. Establish the accuracy of the model needed. It should be useful, feasible, and fit the situation.
13. Test the model and obtain some predicted values. Validate the model by comparing the predicted with true or experimental values. If the difference is too large or statistically significant, work backward to modify or refine the model. If there are no mathematical or logical errors and if the model is less accurate than anticipated, check your assumptions to make certain that they are valid.
14. Always indicate the limits of the model so that the person who uses the model is aware of the restrictions of the model.

   This preceding information is presented for the benefit of biologists who may not be familiar with certain simulation and modeling terminologies. The rest of this chapter is an overview of areas in biology in which computers can be used to manipulate and analyze enormous amounts of complex data.

II. OVERVIEW

A. Computers in Ecology

   There are two principal ways in which computers can be utilized to study ecology. First, computers may be used for data storing, correlation, and statistics. Because ecology deals with such exceedingly complex systems of organisms, it yields data in bewildering richness. Under the best of circumstances, the analysis of this data can be so time consuming as to be practically impossible if one has to analyze it using nothing more sophisticated than, say, a desk calculator. Modern computers have changed this. Now, more and more ecological data are being obtained in a form that can be fed directly into a computer for analysis. The availability of machines that are capable of handling large amounts of data has removed much of the barrier to considering highly complex systems but, needless to say, it has not eliminated the necessity for the researcher to know precisely what he is attempting to learn from his ecological data. On the contrary, the computer has placed more of a burden than before on the ecologist to formulate his questions in a very precise way. One suspects, perhaps, that this secondary effect is at least as important to the study of ecology as the more obvious one of allowing the ecologist to analyze heretofore unassailable quantities of data.

   The analysis of models is the second and, perhaps, more interesting way in which a large computer can be of assistance to the ecologist. Much work has recently been done on the problem of reducing ecological problems to problems in mathematics, that is to say, to the construction of mathematical models. Because problems in ecology are themselves so complex, the models based on the ecological system are also exceedingly complicated. Hence, the analysis of these models has depended in great part on modern computers.

   For example, one might be interested in the fluctuation of a certain population of organisms as a function of the food supply and the predation. At first glance, this might seem quite simple. As the food supply increases, it is certainly reasonable to expect that the population will increase. On the other hand, an increase in the number of predators will tend to decrease the population. Hence, one might obtain a set of mathematical expressions that serve to describe how the population under study varies with the food supply and the number of predators. But even such a simple-sounding model might, in fact, turn out to be quite complicated if one requires, as one should, that the model be capable of predicting results that can actually be observed in the field. For example, an increase in the population of a certain species often heralds an increase in the populations of the species that prey upon the given organisms. It has been documented, for instance, that the populations of snowy owls in the arctic increase during times when lemmings are in abundant supply. And, another consideration: not only is the population dependent upon the food supply, but the food supply per individual is usually dependent upon the population. For example allowing too many sheep to graze on a certain tract of land may kill the grass. And finally, even if the food were in unbounded supply, the territory available to the population is usually not. For instance, it has been observed in some rats that over-crowding can cause a decrease in the population. Hence, constructing a model, which at first glance seemed childishly simple, turns out to be in reality very complex. In the actual construction of any model that is to represent a complicated system, it is safe to say that a considerable period of trial and error ensues before the model begins to predict what can actually be observed. In other words, it is usually necessary to make a number of adjustments in the model before it becomes a reasonable one. Therefore, it is very helpful if one can quickly obtain the effect that varying one of the parameters in a model will have on the model. For this reason, a computer that can, for example, plot the predicted population of a certain species is very useful.

Let us illustrate with an example how modeling can be helpful in predicting the population of a country with some given data. Suppose, it is known from a census data that the population of a particular country has doubled itself in 40 years and we are required to build a mathematical model which can predict a number of years at the end of which time the population will triple. Assume the law of natural growth applies. At time "t", if the population is "N", then:

$$(dN/dt) = k.T; \qquad N = N_0 \, e^{kt} \qquad\qquad (1)$$

$$\text{Here, } t = 40, \, N = 2N_0, \, 2N_0 = N_0.e.40.k; \, e^{40k} = 2 \qquad\qquad (2)$$

When $N = 3N_0$, required to find "t"

$3N_0 = N_0 \, e^{kt}; \, e^{kt} = 3; \, kt = \log_e^3$

Also, $e^{40k} = 2; \qquad 40\,k = \log_e^2$

Hence, $kt/k.40 = \log 3/\log 2; \, t = \dfrac{(1.0986)}{(0.6931)} = \underline{63.2}$ years

If you want to change the assumptions regarding, say, the change in the population of the predator as a function of the change in the population of the organism under study, then you can quickly see the change that the model predicts in the population under study.

B. Computers in Physiology

As in the case of ecology, computers are used in the physiological sciences to analyze experimental data and to evaluate models.

For example, one can extract from data the kinetic constant that relates the rate of reaction to the reactant concentration. Here, the trick is to fit the constant to the experimental data in such a way as to minimize differences between the experimental curve and the appropriate enzyme rate equation. In addition, computers have proved useful in the analysis of results that deal with very complicated metabolic and physiological problems. One such example is the flow of material through branching pathways and across cellular membranes. In many cases, these processes are fairly well understood only on the cellular level. Computers appear to be of great service in extending our understanding to the indescribably more complex case of a multicellular organism. Finally, computers are of great importance in the construction and evaluation of physiological models.

While it is true that any model that can be analyzed on a computer can also, in principle, be analyzed via manual means, the time required for the latter may make it a practical impossibility. And even if it is possible, the time required may be so great as to severely limit the number of alternative models that it is practical to evaluate.

One can quickly appreciate the utility of a computer-borne model by a considerably very simple case of a hypothetical metabolic pathway consisting of several reactions as shown diagrammatically by

$$A \rightleftharpoons B \overset{2}{\rightleftharpoons} C \overset{3}{\rightleftharpoons} \quad \ldots$$

Imagine, furthermore, that each of these individual reactions is well understood, that is, the values for maximum velocity and Michaelis constant, k_m, are known. We simply want to know if the pathway is as drawn, that is, is it a nonbranching series in the order shown? This may be accomplished by comparing the rate of flow through the sequence (which is measurable in the laboratory) with the predicted rate under a variety of conditions. We obtain the predicted rate by solving a set of enzyme rate equations that take into consideration the fact that the product of one reaction is the substrate for the next. We might begin by making the steady-state assumption; i.e., we assume that the concentrations of the inter-mediates remain constant with time. This would require that the rate of formation and breakdown of each intermediate be the same and would therefore greatly simplify things. Under these conditions, we could, perhaps, make our calculation with or without the use of a computer. We would then want to abandon the steady state assumption and introduce, instead, possible branches in the chain or alter the properties of some of the enzymes. One would probably want to alter more than one variable at a time. This would make the use of computer more essential.

Consider the formation of lactic acid according to the following reaction:

$$C_{12}H_{22}O_{11} + H_2O \xrightarrow[\text{lactase}]{\text{enzyme}} 4\ CH_3CHOHCOOH$$
$$\text{milk sugar}$$
$$(A)\quad +\quad (B) \xrightarrow{\hspace{5cm}} (C)$$

For industrial purposes lactic acid is either isolated from sour milk or made by bacterial fermentation. From the point of view of a biologist, lactic acid is very important. It has been called "the keystone of muscular activity". The energy necessary for rapid muscular action appears to be supplied by the decomposition of glycogen to lactic acid. Suppose, we are required to build a model of a bimolecular reaction such as the one shown above, we can begin as follows:

Let us assume, to begin with, there are "a" molecules of "A" and "b" molecules of "B" present in the system. A and B combine to form C. At a time "t" let "N" molecules of C be formed. This is formed by using up N of A and N of B. So, there are $(a - N)$ of A and $(b - N)$ of B left over.

The rate of formation of "C" molecules is proportional to the product of the number of molecules of each substance present.

$$(dN/dt) = k\ (a - N)(b - N)$$

$$\frac{dN}{(a - N)(b - N)} = dt;\ \int \frac{dN}{(a - N)(b - N)} = t + \text{constant}$$

$$t + (E) = 1/(a - b) \int [(1/(b - N) - 1/(a - N)]\ dN$$

$$= \frac{1}{(a - b)} \log \frac{(a - N)}{(b - N)}$$

where T = 0, we have N = 0; (E) = 1/(a − b) log (a/b) substituting for the constant E, we have

$$t + \frac{1}{(a-b)} \log (a/b) = \frac{1}{(a-b)} \log \frac{(a-N)}{(b-N)}$$

$$t = \frac{1}{(a-b)} \log \frac{b(a-N)}{a(b-N)}$$

Likewise, if we try to be so specialized, we could speak on the subject of drug-receptor interaction, and the role of the computer in such a process.

C. Computers in Drug-Receptor Interaction

It was well known that animal cells are bounded by membranes which, apart from anything else, prevent the inner contents from spilling out. The membrane is, however, far more than a passive constraining skin: it is a dynamic structure which controls the passage of chemicals into and out of the cell. Furthermore, it is also responsible for detecting the presence of certain signals such as hormones and for passing on the appropriate message to metabolic machinery of the cell.

Hormones are in the business of communication at the cellular level: they have often been called chemical messengers. They are secreted into the blood stream by specialized glands and exert their effects on particular "target" tissues. Hormone molecules, such as the catecholamines and the protein and glycoprotein hormones have very definite molecular shapes, and their target cells are those which have on the surface of their outer membrane specific receptors for these hormones. Each type of receptor is absolutely specific for binding only one type of hormone. And each cell type in the body has its characteristic hormone receptors. Many hormones, neurotransmitters, drugs, and cellular toxins initiate their action via specific interactions with plasma membrane receptors.

At the present time, the study of hormone receptor mechanisms is entering an exciting phase. Classical receptor theory gives a mathematical description of drug-receptor interaction in terms of a dose-response relationship derived by applying the mass action law to the reversible reaction between the drug molecule and a vacant receptor. This satisfactorily explains most of the experimental findings; however, there are still experimental results which show systematic deviations from the predicted results.

A computer simulation able to check different manners of interaction was tried in order to get similar shapes to the experimental dose-response curves. A dose was taken as a set of random numbers/drug molecules/spread over the elements/cells/ of a matrix/tissue. The criterion for computing the response suggested by the classical receptor theory implicitly supposes that each cell yields a response proportional to its fraction of occupied receptors. However, this might not be the case for some tissues and that is why several alternative criteria for computing a response were classified and used in the computer program which generates dose-response curves. A special attention was paid to the hypothesis of all-or-none functioning cells, having a threshold number of receptors to be occupied in order to onset the release of a "quantum" response. The curves obtained in this case showed a similar shape to the experimental curves for which the classical approach leads to systematic deviations.

When a normal distribution of the threshold of minimal number of occupied receptors was considered, the curves became less steep/the slope for the dose equaling the dissociation constant decreased and the general shape became nearer to the shape predicted by the classical receptor theory. All the generated curves were analyzed by both linear transformations and direct least-squares method. The program is also useful for studying the distribution of drug molecules on the receptors of a tissue under other different criteria: metabolic pathways of the drug, the presence of a competitive or noncompetitive antagonism, the time course of the steady states of a cell, the presence of an endogeneous competition giving a basal response.

D. Computers in Modeling, Structure, and Function of Biomedical Effectors

The investigation of animal limb movements needs the introduction of special models. In high accuracy measurements the acquired data are heavily linked to the model used and the same results obtained in different models may not always be comparable. Numerical calculations in the investigation of biomechanical effectors are very complicated whereas simple topological properties of these systems are very promising in applications. Construction of abstract models for observational purposes in which sets of bones, B, joints, J, and muscles, M, are considered enables the development of comprehensive biomechanical system theory in relational approach. The relational modeling used is a general method for transforming given empirical system in an abstract model. The introduction of R-extremitators as abstract open chains of BJ, BM, or BJM type gives interesting results in the investigation of many biological systems of movement. The description of R-extremitators in generalized fuzzy set notation is useful also in other applications. The determination of suitable functionals on extremitators gives the possibility of comparison of different biomechanical structures, measurement of their similarity, estimation of a coefficient of anthropomorphism for medical modeling, and other biological applications. Topological modeling enables very general and deep insight in the structure and function of biocybernetic systems of movement.

E. Computer-Aided Mathematical Models for the Biological Age of the Rat

The study of influences on the aging process requires mathematical models of the biological age as a standard against which deviations from the so-called "normal age" can be measured. A long-term cohort study with initially 1100 male Sprague-Dawley rats served to establish multiple regression models of biological age and to test influences on aging. Twenty-three parameters from a total number of 42 were selected for a general model.

By means of a factor analysis, the general model was subdivided in 6 factor models of biological age to distinguish between primary and various types of secondary aging changes. Factor 1 can be interpreted as an expression of primary aging. Factors 2 to 5 obviously represent system-specific secondary aging, including compensatory changes. Factor 6 was attributed to general changes in lipid metabolism not directly connected with aging.

F. Computers in Neurophysiological Systems

The mathematical analog-model of the nervous cell is employed to reflect axiomatically the real neuron features. The computer is engaged to investigate the functional activities of the systems composed of such elements. The computer models of random neural nets have been developed with probabilistic-statistical organization in agreement with the available neurophysiological data concerning the central nervous system structures and functions. The initial functional activity of such nets is investigated followed by the subsequent "training" of nets for certain "behavior" types; for that special computer programs are employed. The original training algorithm allows evaluation of the system structural-functioning parameters and to change them according to the instruction goal. Methods of visualizing the above-mentioned computer experiments on the drafting machine BENSON-200 have been developed, including formation of random neuron structures and their dynamics in the training process.

G. Computers in the Determination of the Effect of Body Temperature on Thermal Regulation

There are some trials to deliver breathing gas mixtures at 100% relative humidity at body temperature. To accomplish this requirement and not withstanding the large variability of inspiratory flow patterns of patients requiring mechanical ventilatory support, these devices must incorporate heaters with high power ratings. Heater failures have resulted in fires and thermal injuries to patients.

Four simultaneously obtained temperature measurements on patients receiving mechanical ventilation over an 8-hr period reveal that as the temperature of inspired air is warmed with a humidifier to body temperature, the body heat loss through respiration is reduced. This mechanism requires a redistribution of cardiac output to the skin. Analysis revealed that skin temperature reflects peripheral perfusion. Subsequently, when the inspired gas temperature was decreased the cardiac output and oxygen uptake returned to control values.

H. Computers for Cell Analysis in Hematology

The microscopic inspection process occupies a central role in the hematology laboratory. A major portion of the work in these laboratories is involved with manually locating, classifying, and examining or counting, various cells under the microscope. For example, the purpose of the differential white blood cell examination is to establish the percentage of each of the cell types indicated in the blood stream. This involves manually locating hundreds of blood cells on a stained slide and classifying them into a number of different categories. The percentage of each cell type present is then reported as a result of the examination.

In addition to quantitatively reporting the percentages of different white blood cell types, subjective visual evaluations of the stained red blood cells are also reported. Determinations of typical cell types, variations in red cell shape, variations in red cell size, and estimates of all hemoglobin content are all made. Even though these evaluations are all subjective in nature, they are often critical to the diagnosis of anemia.

These manual processes in hematology are tedious, time-consuming, and sensitive to subjective error. The impact of automation on these visual inspection processes is to relieve the drudgery and improve the speed and throughput of tests performed in the laboratory, while also improving the quality of results. In this regard, the use of digital image processing techniques to analyze and classify peripheral blood cells has developed very rapidly in the last few years. This technology has now matured to the point where there are instruments working routinely in clinical laboratories automatically processing blood slides on a daily basis. An example of a state-of-the-art commercial system for white blood cell classification in routine use today is the Leukocyte Automatic Recognition Computer (LARC).

GENERAL REFERENCES

1. **Coleman, T. E.,** Mathematical modeling as a basis for cardiovascular concepts and experimentation, Proc. of the First Int. Conf. on Mathematical Modeling, Zurich, Switzerland, 1977, p.29.
2. **Coleman, T. E.,** The Concept of Gain in Biological Systems, unpublished material, Department of Physiology, University of Mississippi Medical Center, Jackson Miss.,
3. **Bender, E. E.,** *An Introduction to Mathematical Modeling,* John Wiley & Sons, New York, 1977.
4. **Coleman, T. E.,** Simulation of biological systems: the circulation of blood, *Simulation Today,* No. 51, 1978, p.78.
5. **Beneken, J. E. W.,** A Mathematical Approach to Cardiovascular Function. The Uncontrolled Human System, Institute of Medical Physics, Report No. 2-4-5/6 Ultrecht, The Netherlands, 1965.
6. **Coleman, T. E.,** Simulation in helping biomedical research, *Simulation,* Vol. 19, October 1972, p.29.
7. **Emshoff, J. R. and Sisson, R. L.,** *Design and Use of Computer Simulation Models,* Macmillan, New York, 1970.
8. **Hall, A. D.,** *A Methodology for Systems Engineering,* Van-Nostrand-Reinhold, Princeton, N.J., 1962.
9. **Ackoff, R. L.,** *Scientific Method: Optimizing Applied Research Decisions,* John Wiley & Sons, New York, 1962.

10. **Knuth, D. E.,** *The Art of Computer Programming,* Addison Wesley, Reading, Mass., 1975.
11. **Naylor, T. H., et al.,** *Computer Simulation Techniques,* John Wiley & Sons, New York, 1968.
12. **Ziegler, B. P.,** *Theory of Modeling and Simulation,* John Wiley & Sons, New York, 1976.
13. **Naylor, T. H., Balantfy, J. L., Burdick, D. S., and Chu, K.,** *Computer Simulation Techniques,* John Wiley & Sons, New York, 1968.
14. **Valleron, A. J. M. and Macdonald, P. D. M.,** *Biomathematics and Cell Kinetics,* Elsevier/North Holland, Amsterdam, 1978.
15. **Rosen, R.,** Old trends and new trends in general systems research, *Int. J. Gen. Syst.,* 5, 173, 1979.
16. **Duchting, W.,** Simulation of disturbed cell renewal systems by means of a microprocessor system, *Int. J. Biomed. Comput.,* 10, 375, 1979.
17. **Duchting, W.,** A cell kinetic study of the cancer problem based on the automatic control theory using digital simulation, *J. Cybernet.,* 6, 139, 1978.

Chapter 2

A FOUR-LEVEL SOFTWARE ENGINEERING APPROACH TO MODEL COMPLEX BIOLOGICAL SYSTEMS

S. Sitharama Iyengar, Ramu M. Rao, and Stephen Quave*

TABLE OF CONTENTS

*   The authors would like to express their sincere appreciation to the publishers of the *Journal of Computer Programs in Biomedicine* for permission to include some portions of the authors' original paper, ''A Computer Model for Hydrodynamic Treating of DNA'' in the above-mentioned journal. The authors would also like to thank John Fuller for his ideas on modeling complex systems.

I. INTRODUCTION

Biological systems are very complex, which makes computer simulations and modeling of them difficult and challenging. Furthermore, the physiological complexities of biological systems makes it very difficult to formulate hypotheses to explain their behavior and to test such hypotheses. Many of the most important aspects of behavior of complex systems such as nonlinearity, redundancy, and hysterisis cannot be explained in verbal terms whereas, a mathematical description of these concepts is often both compact and precise. This mathematical description of the system can be described by a logical data structure language. This pseudo language (called an algorithm) can be transformed into any known programming language and can be implemented on any computing system and is also suitable for easy modifications of the model. Mathematical description of the system and software procedure modeling of the mathematical description of the system are playing an increasing role to in understanding the inherent complexity of the system. In this situation, modeling biological systems from a software engineering viewpoint gives the researcher a direct access to formulate a logical structure of all the variables of the system. In the following paragraphs we will attempt to identify those features of software engineering and the modeling process that are most important to biological systems. It is our contention that software engineering techniques are helping to model complex systems, although such a contention is admittedly far from perfect.

The remainder of this chapter is organized as follows: concepts of complexity and computer modeling, software concepts in modeling, an example, and some general conclusions.

II. CHARACTERISTICS OF BIOLOGICAL SYSTEMS

Complex biological systems are those wherein the number of attributes to describe or characterize the systems is too many, and so is the number of variables affecting the system. Not all the attributes are necessarily observable. Very often the characteristics of the biological system defy the definition, philosophy, and scope. In other words, the structure of configuration of the system is rarely self-evident. Over the past 10 years many researchers have been working with various approaches on the development of models of biological systems such as growth of cancer cells, pharmocological activity of a particular drug on humans and animals, understanding and working of DNA molecules, and cognitive process of human systems. Most notable are the works of Crick,[3] Davidson and Britten,[1] Leventhal and Davison,[12] Guyton and Coleman,[20] Coleman,[24] Iyengar and Quave,[21,22] and Iyengar.[23] Recent studies of modeling complex biological systems are available in the following references: 25, 26, 27, 28, 29, and 30.

A. Modeling Criteria of Biosystem

There are a number of criteria that must be considered in choosing a model for a given system. These include generality, development of an algorithm, computational effort, storage requirement, numerical stability, and implementation effort. Computational effort is tied to the storage requirement since one may choose to recompute values to save storage. We may think of the computational effort associated with the type of statistical modeling used. It is difficult to generalize about storage requirements since the storage requirements are dependent on both the problem solved and implementation. For most models which are used in practice, all of the algorithms are fairly stable. However, for some parameter values, some or all of the algorithms will experience numerical difficulties.

B. Model Behavior and Complexity

The use of models of computation allows the examination of the behavior of systems we wish to study and focuses attention on existence and complexity processes for the model.

Software analysis technique simulates the behavior of what we wish to investigate, abstractly yet precisely. Thus, the desired characteristics of models are classified as follows:

1. Accuracy of the estimated parameters
2. Number of parameters of the model
3. Whiteness of model residues

A detailed analysis of these characteristics is given in the paper of Maklad and Nichols.[25] Now, we shall describe briefly the "importance aspect" of the model building process that has been generally neglected by many model builders. The practical impact of this characteristic in model structure discrimination is as follows. As the number of experimental data used for estimation increases, the basic parameter estimates of an accurately estimated model will not change appreciably. Those of inaccurately estimated models will. Maklad and Nichols[25] state that "if we do not include the accuracy of the parameter estimates in the model discrimination criterion, the choice of a model will be sensitive to available data".

C. Number of Parameters of the Model

If a number of models satisfy the accuracy of the estimated parameter and whiteness of model residues then it is wise to choose a model with the smallest number of parameters. Furthermore, the criterion in choosing a model with the smallest number of parameters is based on the complexity of the models.[27]

D. Whiteness of Model Residues

A good model should be able to predict the behavior of a system under wide operating conditions. Therefore, it is natural to require innovations to be as close as possible to white noise. For more on this refer to the paper by Maklad and Nichols.[25]

The development of a computer model for a biological system requires various simplifying assumptions. Especially when performing a theoretical analysis, they are necessary in order to limit complexity to a bearable dimension.[26] In this context, a comment on the complexity of the system seems in order. Complexity has been defined as "a measure of the difference between a whole and the one-interacting composition of its components". Figure 1 shows a system repeatedly subdivided into levels of increasingly smaller subsystems until at the bottom level there are only the simplest components, v_1, v_2, . . . , v_n, which can be decomposed no further. Maklad and Nichols,[25] in addition, give the complexity $C_1(S)$ of a system S by the following formula:

$$
\begin{aligned}
C_1(S) &= C_1(S_1) + C_1(S_2) + r(S_1, S_2) \\
&= C_1(S_{11}) + C_1(S_{12}) + R(S_{11}, S_{12}) + C_1(S_{21}) + C_1(S_{22}) + \\
&= C_1(S_{23}) + R(S_{21}, S_{22}, S_{23}) + R(S_1, S_2) \\
&= \ldots \\
&= R(S_1, S_2) + R(R_{11}, R_{12}) + R(S_{21}, S_{22}, S_{23}) + \ldots
\end{aligned}
$$

where R(.,..,....,.) signifies interaction between the arguments and where the sum of the complexities $C_i(V_i)$ where i varies from 1 to n, is 0. The result is that the complexity of the system, S, has been reduced to the sum of the interactions which now only need to be defined. We refer to the above formulation as Maklad-Nichols Solution.

In terms of the concepts of models presented above, a system can be decomposed in a manner similar to that proposed by Maklad and Nichols[25] until the lowest level consists of entity occurrences only. It may be helpful in modeling some systems to allow the most basic interactions between these simple components. For example, the number of variables in a cardiovascular model are too many. Therefore, identification of parameters can be done by breaking the cardiovascular model systems into subsystems.

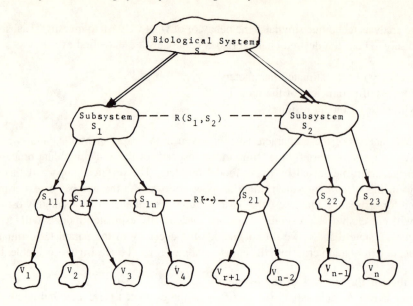

FIGURE 1.   General decomposition of a system.[25]

The subsystem at the next highest level, to which the simple components belong, will generally be an entity type (used for a specific purpose) with interactions, or relationships, with other entity types. The next level could be considered the data structures to represent these entity types and the interactions between these data structures that represent the relationships between the entity types. There are, perhaps, additional interactions due to the nature of the data structures rather than to the relationships between the entity types. Thus, in moving from one level to the next higher level, the interactions between the subsystems are becoming increasingly complex. Furthermore, it is now possible to define these interactions in "algorithmic" terms. The next level may consist of subsystems formed from modules or groups of modules of the total algorithm. This may in fact be repeated over several levels with each subsystem of a higher level being related to subsystems of the previous levels. From our data structure viewpoint, the final level will be the program that actually implements the system. Diagram 1 helps us to identify the process of estimating four parameters of a given biological model system.*

Note that the desirable property of a model is related to both parameter estimation and model residues. For more on this refer to the paper by Maklad and Nichols.[25]

III. DATA STRUCTURE + ALGORITHM + MATHEMATICS =
COMPUTER MODELING OF SYSTEMS

Computer modeling of biological systems is a process of building a new set of conceptualizations, theories, and implementations. Furthermore, it encompasses three distinct areas of science (computer science, biology, and medicine) for building a framework for the investigation of the behavior of a complex biological system.

In general, the study of computer modeling of systems is primarily concerned with the study of data representation (or data structures, as they are often called) and their transformation to an empirical model by a software program. The data manipulated by a program provides descriptions of objects in the real world. We are interested in the accuracy and relevance of these descriptions. Because the computer model seeks to represent difficult real world problems with suitable data structures and algorithms for varying observational data,

*   We acknowledge the assistance of John Fuller in this project.

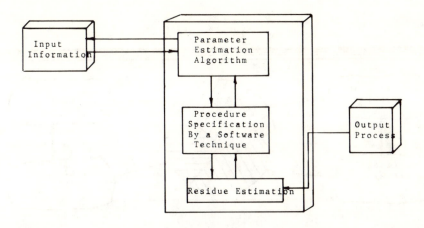

DIAGRAM 1.

there are many issues associated with the representation process of the system model. A central basic issue in modeling a complex biological system (such as growth of cancer cells) is to understand the relationships between the data elements that are relevant in describing the properties of the system (such as continuity, simultaneity, and interdependency). Also, it is very important to develop efficient statistical algorithms to fit data elements to a great number of different distribution functions or nonlinear relationships. Since electronic computers are widely used in statistical analysis, it is essential that the model developer be aware of the following four levels in the representation and programming process of the model (Figure 2).

1. Representation Level (RL): This level consists of the representation of the observed data by a suitable data structure.

2. Functional Level (FL): The functional specification is an implicit characterization of the structure at and by functions and properties of these functions. In other words, the functional dependencies between data elements in the modeling process must be observed.

3. Algorithmic Level (AL): The logical description of data elements provides a means to construct the structure with a set of base objects and constructors. Algorithms for the manipulation of data elements representation (data structure) must be efficient. Since the data objects and constructors are purely mathematical entities, we may consider the algorithmic level as a general logical framework for expressing logical descriptions of the modeling process and hence not dependent on any computing machine. Therefore, the algorithmic level is a model framework describing the logical relationship of all the performance variables of the system.

4. Tools Level (TL): For modeling complex biological systems, the modeling tools have to be carefully selected based on a thorough understanding of the essential hardware and software requirements of the system.

We are now ready to describe the software aspects of the modeling system.

COMPUTER MODELING
OF COMPLEX SYSTEM

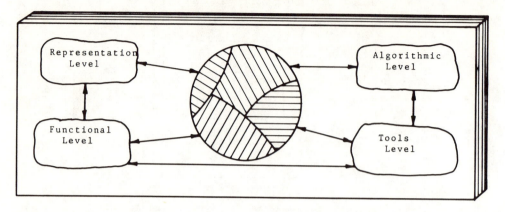

FIGURE 2.   Four level framework in modeling complex systems.

IV. SOFTWARE POINTS IN MODELING OF SYSTEMS

Software engineering tools are now being proposed as a way to solve problems of modeling complex biosystems such as high costs and hidden errors. These techniques have tremendous potential to structure the modeling process, making it more reliable and error free. The potential also exists in representing interaction of parameters precisely associated with any complex system. Topics of interest include (1) the design of a suitable data structure and (2) development of computationally efficient algorithms for a data structure model. The purpose of this paper is to discover the forces surrounding the adoption and use of software engineering techniques to model complex biological systems. A software environment should provide the following:

1.   A broad interface between data models and programming languages; the data and the control structure of the language should interact smoothly with the data model algorithms.
2.   In an environment of modeling biosystems, there is a high demand for debugging aids, statistical package programs, and error-handling systems.

Many programming languages provide the facility of grouping several related variables of the system into a logical group, usually called a logical data structure. Data structures and algorithms occur at two different levels. The model developer in a planning stage thinks of the data structures and algorithms on a conceptual level. Thus, we may regard data modeling as a successive refinement of the following two quantities: (1) data collection of the existing system and (2) data analysis. Data analysis is a procedure for developing a data model by examining systems and data in separate areas of organizations. Thus, a logical data structure is generated and hence can be optimized for response to a nonlinear type of datum which is normally true in biological systems. The logical model of the required data (produced by information or data analysis) is used as an input for software procedure modeling of the system. For more on this, refer to any software engineering book and *Computer World* magazine on data modeling.

Thus, the software procedure modeling of the biological systems may be regarded as consisting of the following steps:

Step 1.   Program specification synthesis — this is a methodology for formal derivation of program logic directly from a data model produced by data analysis.

Step 2.   Breaking the derived program logic into blocks of structured program statements giving a high level algorithm for performance of the system.

Step 3.   Compilation of resultant logical model (obtained in Step 2) to the programming language level (one that can execute on a machine).

Step 4.   Software evaluation of the system.

This method of modeling complex biological systems automatically exhibits high functional cohesiveness of the nonlinear data behavior of the system. With this background in mind, we shall now describe an example to model a complex biological system.

V. EXAMPLE — COMPUTER SIMULATION OF HYDRODYNAMIC SHEARING OF DNA BIOSYSTEM

A. Introduction

The DNA molecule contains an immense amount of information in coded form. This coded information controls the growth, development, and major characteristics of biological systems. The control of the information stored in DNA is a major problem of research interest in molecular biology.[1] The first level at which control can be exerted is called "transcription". At this level, segments of the linear DNA base sequences are transcribed into linear sequences called "messenger ribonucleic acid", (mRNA). In this process, the linear sequences of the four symbols, A.T.G. and C. which is the alphabet of the code in DNA is transcribed into a new code in mRNA which has the symbols A.V.G., and C. The transcription is one to one with the following correspondence: A into V, T into A, C into G, and G into C. These sequences which are part of the whole DNA molecule are transcribed in units called genes. The mRNA is translated into proteins by a process known as "protein synthesis". There are many hypotheses concerning mechanisms of transcriptional controls. Some or all of these mechanisms may be operating but none satisfactorily explain all observed phenomena. Many of the recently proposed control mechanisms (1, 2, 3, 4, 5, 6) involve the organization of the genes along the DNA strand. These proposals give varying functions of sequences which are found repeated many times along the DNA strand. These repeated DNA sequences were developed at Carnegie Institute by Britten and Kohne.[7] The rehybridization process of the system can be explained as follows:

1.   The very long DNA strands have been shortened by hydrodynamic shearing
2.   The double strands are separated by heat, the strands become (uniformly) randomly distributed in solution
3.   Upon lowering of the temperature, the single strands rehybridize into double strands by a random collision process

B. Statement of the Problem

The purpose of the study is to propose a distribution function for the pattern of breaks along a DNA strand after hydrodynamic shearing. This distribution is determined by building a computer simulation model validated by reproducing or being consistent with real world experimental evidence and capable of giving break patterns by appropriate simulation runs.

C. Application of Four-Level Approach

1. Representation Level

Physical model — The conceptual model for DNA undergoing hydrodynamic shearing of DNA is that of blocks with stiff but flexible connecting elements which can be broken. These strands are suspended in aqueous solution which is undergoing Brownian motion and mass motion due to stirring.

Stochastic description of the system — When forces are being applied to the connecting elements of the model, the blocks undergo finite but complex vibrational motion and have complex patterns of vibration. As a consequence, the connecting elements may be stretched or compressed. Therefore, one cannot predict exactly what force will be required to break any given connector, but there will be an expected value and a distribution function that together describe the force that will break connectors. The distribution of strands in solution is assumed to be uniformly random. The strength of connector elements to which the shear force is applied is a stochastic variable whose distribution function is unknown, but which has a strong central tendency.

2. Functional Level

This section of our chapter describes the formation of the model using mathematical data structure.

When a force is applied to a connector, it can break or not break depending upon the following factors: (1) velocity gradient — a function of unknown distribution, (2) the position chosen on the strand — an unknown density function with an extreme central tendency, (3) the strand length — which can be modeled as a stochastic difference equation, and (4) the position and orientation of the DNA chain in solution — a uniform random variable, but whether it breaks or not is a Bernoulli's trial.

Therefore, one can see that the number of strands that break out of N such strands is a binomial process, but the probability of a break is a complex interaction of a function of several stochastic variables some of which have unknown distribution functions.

The central limit theorem allows one to approximate the random contribution from several populations each with the same or different underlying distributions. A Gaussian distribution function will, therefore, be used to choose the connector element where the stress occurs. This connector interval will be drawn from a N(length/2, sigma) population, where sigma is on the order of 0.04 to 0.05.

It will be assumed that when the shear forces first start and for at least a short period thereafter every strand in solution undergoes a trial in period t, the time for the average strand length to decrease by one half. This assumption is certainly true for the real world experiment during the first few time periods.[12] To be precise:

$$\begin{aligned} &\text{average molecular weight or length (t)} = \\ &2^{-Kt} \text{ (starting molecular weight)} \\ &\text{for t} = 0, 1, 2, 3, 4, \text{ or } 5 \end{aligned} \qquad (1)$$

Let us assume that one starts with N strands. Within a time t, all strands undergo a trial. After each trial, a strand can be broken or not; that is, each strand undergoes a Bernouli trial, and each trial is independent of any other. Therefore, the number of strands out of N that will break in time t, is a binomial random variable. In particular:

$$\begin{aligned} &\text{the probability of K breaks in N trials} \\ &\text{during time period t is P(K,N) and P(K,N)} = \\ &\hat{P}_t \, K_{(1-\hat{P}t)} N - K \\ &\text{where } \hat{P} \text{ represents p hat} \end{aligned} \qquad (2)$$

Here p hat is used because the probability of a break is not constant for each strand undergoing trial. P_t is a function of the position where stress is applied, a random variable. Let LEN represent the length of the short side of any potential break point.

$$P_t = f(LEN) \qquad (3)$$

for each strand tried. This function of LEN will have an average or expectation value for all strands in period t, or:

$$P_t = f \, E(LEN) \tag{4}$$

If L is the length of all N strands at t = 0 then

$$P_1 = (L/2) \tag{5}$$
$$\text{Where } L/2 = E(LEN) \tag{6}$$

If the N strands are unequal in length (at t = w and thereafter), then

$$P_t = E(LENGTH_{t-1}/2) \tag{7}$$

or a function of one half the average length from the previous time period. Ideally, P_t should also be a function of the velocity gradient G (which in turbulent flow is a random variable of an unknown density function) and the force necessary to break the connector element where force is applied (either a Maxwell density function or an empirical distribution function). However, to avoid complexity, one can use Equation 10 and neglect other processes. One can neglect the variation in force necessary to break a connector element, and the velocity gradient is a function of the stirring speed in the blender or rotor. That relation is

$$\text{average base pair length} = E(LENGTH) =$$
$$13186570.05/X + 452.1452331 \tag{8}$$

where X is the rotor speed. The linear correlation coefficient for this equation is 0.983.[10] Therefore, one could use a parameter LAMBDA which could be related to rotor speed by LAMBDA = MX + C. The probability, P_t will be a function of the random variable LEN and the parameter LAMBDA as an approximation in the simulation model. The function that is used for P_t is arbitrary to some degree. This function is the cumulative exponential distribution:

$$P_t = 1.0 - EXP(-E(LEN)/LAMBDA)^M \tag{9}$$

Here M is arbitrary and controls the variance of the distribution of lengths. LEN will be drawn from a normal population, N(LENGTH/2,), where LENGTH is the average molecular weight or strand length from time period t − 1. Or,

$$LEN = \text{sample from } N(0,1) * SIG * LENGTH/2 + LENGTH/2 \tag{10}$$

$$E(LEN)_t = LENGTH_t - 1/2. \tag{11}$$

Using Equations 2, 7, and 5

$$P[K(t), N(t-1)] = \binom{N(t-1)}{K(t)} [1.0 - EXP(-LENGTH(t-1)/2*LAMBDA)^M]^{K(t)}$$

$$*EXP[(-N(t-1)*LENGTH/2*LAMBDA)^M]^{(N(t-1)-K(t))} \tag{12}$$

Where K(t) is the number of breaks in period t, $N(t-1)$ is the number of fragments from the previous time period, and $LENGTH_{t-1}$ is the average fragment length from the previous time period. At the end of time period t, there will be N(t) fragments where

$$N(t) = N(t-1) + K(t) \tag{13}$$

Thus, after having established the functional characteristics of the system, now we proceed to describe an algorithm which simulates the behavior of the shearing process of DNA.

3. Algorithmic Level
The shear stress model embodies the following algorithm (major blocks of processing).

Step 1.   Choosing the stress point, the site of the event
Step 2.   Finding the probability of occurrence of a break at the stress point (probability of an event process)
Step 3.   Executing a Bernoulli trial, deciding whether a break occurs
Step 4.   Storing a temporary event vector for use in the time-slice proceeding
Step 5.   Updating the event vector for time-slice 0
Step 6.   Collecting breaking statistics for each strand
Step 7.   Summary of statistics for sensitivity analysis

We are now ready to describe the tools level of the modeling process of the system. (See Figure 2 for a flow chart description of the algorithmic process.)

4. Tools Level
This level of the modeling essentially describes the software and hardware requirements of the modeling process of the system. The computer software for the simulator is as follows:

A.   Listing
B.   Description
    1.   Main program
        a.   Input
        b.   Loop structure
            1.   Outer loop and parameters
            2.   First inner loop, the processing of each strand and strand parameters
            3.   Second inner loop and time slices
            4.   Innermost loop, within time slice processing
        c.   Processes
            1.   Choosing stress point
            2.   Finding probability of break
            3.   Break decision
            4.   Storing events during time slice
            5.   Updating event calendar for each time slice
            6.   Collecting statistics for validation and run purpose
    2.   Subroutines
        a.   Generation of standard normal deviate
        b.   Generation of uniform random deviate

a. Hardware Requirements
Modeling of biological systems can be expensive if not planned ahead with regard to the type of hardware to be used. The hardware architecture of the computing system can be a

major factor in the design of a simulator for a biosystem. The simulator which simulates the shearing process of DNA was implemented on an IBM computing machine. The generality of formulation provides an opportunity for increasing or decreasing the completeness of system representation, a distinct advantage in dealing with a computer system.

b. Software Description

The input consists of tables and starting parameter values, exogenous variables. The first input is NG0, the number of iterations of the outer loop. The NBEGN1, NBEGN2, NB(1), SIG, RLAMDA are read in. NBEGN1 and NBEGN2 are starting integers for UNIRN1 the first uniform random number generator. Any large odd integers will do, and different pairs will initiate a new sequence of random numbers. NB(1) is the starting fragment length for all strands. SIG is a parameter in the Z transformation of N(0,1) in choosing the stress point; that is, stress point = (SAMPLE FROM N(0,1)*SIG*(FRAGMENT LENGTH/2) + FRAGMENT LENGTH/2. RLAMDA is LAMBDA the parameter which controls the average molecular weight (strand length). Next the table from the National Bureau of Standards for cumulative standard normal distribution is read into Y and X. Then, NBEGN3 and NBEGN4 are read in; these variables serve as starting integers for UNIRND2, a second uniform random number generator. Following this, MINLEN and MAXLEN are input. These values are the minimum and maximum values for the next table. Next the table itself, is input in the vector LENN. Which is an ordered event list consisting of a selected number of possible values of LEN, an endogenous stochastic variable. Then the probability of occurrence associated with each value of LENN is input into the vector RLENN. MINLEN and MAXLEN are chosen such that the probability of occurrence is near zero and one respectively. These probabilities are precalculated using P(EVENT LEN) = $1.0 - EXP(-LEN/LAMDA)^M$. M being chosen to control the variance of the event list LENN and, thus, the variance of distribution of lengths.

The loop structure of the program is informative. The outermost loop is DO 999 III = 1,NG0. This outer loop is used for reinitializing NBEGN1, NBEGN2, NBEGN3, and NBEGN4. These input variables control the sequence of uniform random numbers from subroutines UNIRND1 and UNIRND2. If one wishes to get a series of runs each using the same series of random numbers, NG0 is chosen appropriately. KK, LL, L1, K1 are also chosen to reinitialize tables within the subroutines as will be explained when these subroutines are discussed in detail. The next innermost loop is DO 880 JKL = 1, 1000. This loop controls the number of strands to be processed. DO 309 IPI = 1,15 is the next most inner loop, and it controls the number of time slices each strand is processed. The innermost loop is GO TO 020 and is taken as long as the number of fragments tried for breakage within each time slice is less than the number generated within the last time period i.e., is J.GE.JO in which case GO TO 020 does not exercise control, and processing for the next time slice is initiated.

The major blocks of processing can be seen in the processing flow chart. They are

1. Choosing the stress point, the event
2. Finding the probability of occurrence of breaks at stress points, probability of event
3. Executing a Bernoulli trial, deciding if a break occurs
4. Storing temporary event vector within for time slice processing
5. Updating event vector for time slice, t
6. Collecting and printing events and statistics for each strand
7. Printing summary for all strands

Throughout the program NSENSE is used as a switch. If NSENSE is off, i.e., if NSENSE = 0 then the vector NEWNB, contains the state of the event space, the list of break points for all previous time slices for a strand. If NSENSE = 1, then NB contains the state of the event space and is used for processing. Although this decision switch complicates programming, it saves a storage-to-storage transfer of NEWNB to NB at the end of the time slice.

Choosing the stress point starts by generating a uniform random variable by calling UNIRN1 (which will be discussed in detail below). A decision is then made as to whether RTVAL is less than or greater than 0.5. This is done because subroutine RNORML accepts values between 0.5 and 0.9999997 only. NP is used as a switch, i.e., NP = 0 if RTVAL is less than 0.5, and RTVAL = 1.0 − RTVAL. RTVAL is then sent to RNORML and a standard normal deviate is returned. The details of RNORML are discussed below. The standard normal random variable is returned in RTRVAL. If NP = 0, RTVAL is made negative. The stress point NX is then calculated by NX = RTVAX*RMEDIN + RMEDIN (see Equation 10).

Once the stress point is chosen, the probability of a break at that stress point is a function of LEN, the length of the short side of a chosen stress point. This probability P = 1.0 − EXP − (LEN/RLAMD)^M may be calculated as needed, but it is much faster to precalculate P for MINLEN to MAXLEN and store in a vector. Here, PNX is found in the address (LEN − MINLEN + 1), and execution time is significantly decreased. If LEN is greater than MAXLEN, a break always occurs; and if LEN is less than MINLEN, a break never occurs.

1. Software Description for the Generation of Uniform Random Numbers

The method of choice for generating pseudorandom numbers is the linear congruential difference (Equation 13).

$$I_{N = IN-1}*M \text{ (MODULUS } 2^W) \tag{14}$$

where I_{N-1} is the previous integer in the series, M is a multiplier, and W is the word size of the computer being used.

This method passes many of the tests used for uniform random distribution. Although many other methods have been proposed[14,15] the purely multiplicative generator, Equation 14, stands up well under the most stringent test for uniform random numbers if the word length is of sufficient size.[14,16] In this test based on Fourier analysis, it was found that machines with word sizes of less than 35 do not give good results on the spectral test.[16]

Because the IBM SYSTEM has a word length of 31, it was decided another method should be used. However, MacLaren and Marsaglia[17] have developed a method based on the combination of two linear congruential generators, and their algorithm passes the most stringent test even when used on a machine with a relatively short word.[13] Moreover, the choice of M, the multiplier, is usually critical, but by using the combination method of the two multipliers, M_1 and M_2, one escapes this stringent requirement. Even two poor choices will give a series with good statistical properties.[13,17]

The method of MacLaren and Marsaglia was used in UNIRN1 and UNIRN2. Two generators are used; one each for the two stochastic processes of choosing a trial event and the Bernoulli trial for event occurrence. By choosing four different multipliers, the two generators are made statistically independent; moreover, the runs are more reproducible when two runs use the same integer starters and therefore the same series of random numbers.

The algorithm in both generators is the same. First, a table of 128 integers is generated by the first linear congruential difference equation. The passed parameters LL and KK for UNINRN1 and L1 and K1 for UNIRN2 control the creation of the table by acting as switches. If LL and L1 are zero, the tables are created. After the tables are created, these values are made nonzero, and the table is not created again unless the switch is turned off. Next, a 129th integer is produced and stored temporarily. The second linear congruence difference equation produces a number between 1 and 128 by dividing I_N by 2^{24} and adding 1. I_N can have any value between 0 and $2^{31} - 1$. Consequently, 1 of the 128 values is chosen using the random integer produced by the second difference equation. This integer chosen from the table is then multiplied by the real equivalent of $1/(2^{31} - 1)$ which results in a number in the range 0 to 1. The 129th integer generated by the first difference equation is used to replace the value chosen randomly from the table by the second difference equation. This algorithm produces pseudorandom series with very good statistical properties, and the execution time after the table is generated is only slightly more than that required for a single equation generator.[13]

2. Software Description for the Generation of Standard Normal Random Numbers

The simplest way to produce a random deviate with a nonuniform distribution (X) is to first produce a uniform random deviate R and then to find the inverse $F^{-1}(R)$ of the nonuniform cumulative distribution function. However, the inverse

$$X = F^{-1}(R) \tag{15}$$

of the cumulative distribution function does not exist for the normal distribution, but tables for selected values of X can be generated from:

$$F(X) = \int_{-\infty}^{x} \frac{1}{2\pi} e^{-1/2x^2} \, dx \tag{16}$$

Using this table one could generate X the random normal deviate by generating RTVAL a uniform random deviate between 0 and 1, converting RTVAL to a number between 0.5 and 0.9999997 if RTVAL was less than 0.5, setting RTVAL F(X) in Equation 16, finding F(X) just larger than RTVAL in the table, using linear interpolation to find X, and multiplying X by −1 if RTVAL was less than 0.5. This sequence is used in subroutine RNORML. The search for the slot within which RTVAL falls takes advantage of the statistical nature of the list. The 192 probabilities in the list (a National Bureau of Standards cumulative standard normal distribution table) is first divided into 6 sublists of 32 items. The last item in each of the 6 sublists is compared to GAUSS which is RTVAL from the main routine. In this manner, one finds the sublist of 32 within which GAUSS falls. The statistical nature of the list is such that GAUSS will fail within the first sublist 46% of the time, the second 72%, the third 92%, the fourth 98% of the time, etc. After the first sublist of 32 is found, it is divided into 4 further sublists of 8 items each. When GAUSS is found in an 8 sublist, this further sublist is divided into 4 sublists of 2 items each. One more comparison decides the slot within which GAUSS falls. Next linear interpolation is used to transform RTVAL into GAUSS a standard normal deviate. GAUSS is returned to RTVAL in the main program and is made negative if RTVAL from UNIRN1 was less than 0.5.

D. Results of the Model

The following are experimental results of simulation model and comparison to real world data, the model developed is of the following form:

Model 1: P_B = 1.0 − 1/Exp (LEN (LAMBDA))
Model 2: P_B = 1.0 − 1.0/Exp ((LEN/LAMBDA)^2)

Model 1 was the first model used, but a completely debugged simulator was never produced using model 1. However, in preliminary and partial results, it was seen that it would take too much time, "t", to process a single strand. Although model 2 does not give a valid

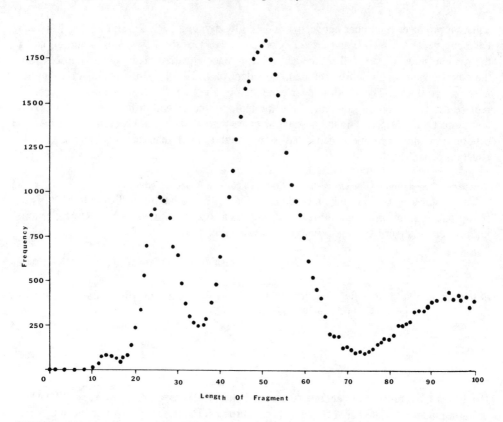

FIGURE 3.   Under the conditions: LAMBDA = 50, SIG = 0.05, and 16 time periods of simulation, the output of breaks from MODEL.

distribution of break length for the end product of the blender shearing experiment, it does give results indicative of the real experiment at an intermediate stage. For more on the behavior of these models, refer to papers by Iyengar and Quave.[21-23] Figure 3 describes the distribution of breaks from model 2. The parameter values used for these conditions are PSIG − 0.05 and LAMBDA = 80. In this simulation, the number of time periods used was dependent on the relative change in the number of strands.

E. Conclusions

The software engineering approach provides an opportunity for increasing or decreasing the completeness of system representation, a distinct advantage in dealing with a complex biological system like DNA. Furthermore this method will provide an instrument for the design of meaningful experiments and parametric investigation on the performance of the system. As in other sciences, exploration of the modeling process occurs at several levels of analysis. The framework presented in this paper provides a general perspective on conceptual modeling of a complex biological system based on four different levels of data abstraction and data analysis.

REFERENCES

1. **Davidson, E. H. and Britten, R. J.**, Organization, transcription, and regulation in the animal genome, *Q. Rev. Biol.,* 48, 565, 1973.
2. **Bonner, J. and Jung-Rung, Wu,** A proposal for the structure of the *Drosophila* genome, *Proc. Natl. Acad. Sci. U.S.A.,* 70, 535, 1973.
3. **Crick, F.,** General model for the chromosomes of higher organisms, *Nature (London),* 234, 25, 1971.
4. **Georgiev, G. P.,** On the structural organization of operon and the regulation of RNA synthesis in animal cells, *J. Theoret. Biol.,* 25, 473, 1969.
5. **Georgiev, G. P.,** The structure of transcriptional units in eukaryotic cells, *Curr. Top. Dev. Biol.,* 7, 1, 1972.
6. **Jelinek, W., et al.,** Further evidence on the nuclear origin and transfer to the cytoplasm of poly(A) sequences in mammalian RNA, *J. Molec. Biol.,* 75, 515, 1973.
7. **Britten, R. J. and Kohne, D. E.,** Repeated sequences in DNA, *Science,* 161, 529, 1968.
8. **Botchan, M. R.,** Bovine satellite I DNA consists of repetitive units 1,400 base pairs in length, *Nature (London),* 251, 288, 1974.
9. **Mowbray, S. L. and Landy, A.,** Generation of specific repeated fragments of eucaryotic DNA, *Proc. Natl. Acad. Sci. U.S.A.,* 71, 1920, 1974.
10. **Davidson, E. H., Hough, B. R., Amenson, C. S., and Britten, R. J.,** General interspersion of repetitive with non-repetitive sequence elements in the DNA of *Xonopus, J. Molec. Biol.,* 77, 1, 1973.
11. **Bresler, S. E.,** *Introduction to Molecular Biology,* Academic Press, New York, 1971.
12. **Leventhal, C. and Davison, P. F.,** Degradation of deoxyribonucleic acid under shearing forces, *J. Molec. Biol.,* 3, 674, 1961.
13. **Knuth, D. E.,** *The Art of Computer Programming,* Vol. 2, Addison-Wesley, Reading, Mass., 1969.
14. **Lewis, T. G. and Payne, W. H.,** Generalized feedback shift register pseudorandom number algorithm, *J. Assoc. Comput. Mach.,* 20, 456, 1973.
15. **Westlake, W. J.,** A uniform random number generator based on the combination of two congruential generators, *J. Assoc. Comput. Mach.,* 14, 337, 1967.
16. **Coveyou, R. R. and MacPherson, R. D.,** Fourier analysis of uniform random number generators, *J. Assoc. Comput. Mach.,* 14, 100, 1967.
17. **MacLaren, M. D. and Marsaglia, G.,** Uniform random number generators, *J. Assoc. Comput. Mach.,* 12, 83, 1965.
18. **Skala, A.,** A method for the breakage of DNA and resolution of the fragments, in *Methods in Enzymology,* Vol. 11, Grossman, L. and Moldave, K., Eds., Academic Press, New York, 1971.
19. **Bernhard, S.,** *The Structure and Function of Enzymes,* W. A. Benjamin, New York, 1968.
20. **Guyton, A. C. and Coleman, T. G.,** Quantitative analysis of the pathophysiology of hypertension, *Circ. Res.,* 24(1), 1, 1969.
21. **Iyengar, S. S. and Quave, S. A.,** A computer model for hydrodynamic shearing of DNA, *J. Comput. Progr. Biomed.,* 9, 160, 1979.
22. **Iyengar, S. S. and Quave, S. A.,** A computer model for hydrodynamic shearing of DNA — part II, *J. Comput. Progr. Biomed.,* 10, 133, 1979.
23. **Iyengar, S. S.,** A computer model for hydrodynamic shearing of DNA — part III, *J. Comput. Progr. Biomed.,* May 1981, Vol. 11.
24. **Coleman, T. G.,** Simulation is helping biomedical research, *Simulation Today,* No. 8, 29, 1976.
25. **Maklad, M. S. and Nichols, S. T.,** A new approach to model structure descrimination, *IEEE Trans. Syst. Man and Cybernet.,* Vol. SCM-10, No. 2, February 1980.
26. **Fasol, K. H. and Jorgl, H. P.,** Principles of model building and identification, *J. Automatica,* 16, 505, 1980.
27. **Van Eden,** *An Analysis of Mathematical Complexity, Mathematical Centre Traits,* Vol. 35, Mathematical Centrum, Amsterdam, 1971.
28. **Astrom, K. J.,** Maximum likelihood and prediction error methods, *J. Automatica,* 16, 551, 1980.
29. **Anderson, B. D. O., Moore, J. B., and Hawkes, R. M.,** Model approximation via prediction error identification, *J. Automatica,* 14, 615, 1978.
30. **Coleman, T. G.,** The concept of gain in biological system, Proc. of the First Int. Conf. Mathematical Modeling, July 1, 1977.

Chapter 3

APPLICATION OF STATISTICAL TECHNIQUES IN MODELING OF COMPLEX SYSTEMS

Musti S. Rao and S. Sitharama Iyengar*

TABLE OF CONTENTS

*   The authors would like to express their sincere gratefulness to Dr. Andrew P. Sage, editor, *IEEE Transactions on Systems, Man and Cybernetics* for permission to include some portions of the authors' original paper, "Statistical Techniques in Modeling of Complex Systems; Single and Multiresponse Models" published in the above-mentioned journal.

I. INTRODUCTION

A. Mechanisms and Models

The basic aim of the study of a chemical, physical, or biological system is to investigate how it behaves so as to make recommendations for its future developments. The most common means of doing this is to build a mathematical model for the system through which it becomes possible to predict, control, and optimize the system.

For any system, it may be assumed that there exists a precise mathematical and physical representation of all phenomena that make up the system. In many situations, however, *a priori* it may not be known what this model is, and in fact, the first goal of the experimenter is to obtain this relationship which is normally called the model.

For example, the biological oxygen demand (BOD), which is used as a measure of the pollution produced by domestic and industrial wastes, may be given by an exponential model of the form[1]

$$\eta = \theta_1 \, (1 - \exp - \theta_2 t) \tag{1}$$

where η is the BOD and t is the incubation period.

Another example of a model is the expression for the rate of a chemical reaction between two species A and B, which may be represented by

$$r_a = - dC_a/dt = k \, C_a^m \, C_b^n \tag{2}$$

where: r_a is the rate of consumption of the species A, k is the rate constant of the reaction, C_a and C_b are the concentrations of the species A and B, t is time, and m and n are the reaction orders. Equation 2 is a power-law model in which the reaction rate r_a is the dependent variable and the concentrations C_a and C_b are the controllable quantities.

Normally, in practical situations, one encounters more complicated models. However, in general, all phenomena can be theoretically represented by a mathematical model of the form

$$\eta = f \, (\underline{\theta}), \, (\underline{\xi}) \tag{3}$$

where: η is the expected value of the dependent variable y, $\underline{\theta}$ is a (px1) vector of parameters, and $\underline{\xi}$ is a (px1) vector of independent variables.

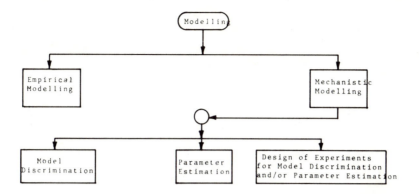

FIGURE 1. Modeling strategies of systems.

It is frequently possible to postulate several physically meaningful models which can describe the same system. The investigator is then faced with the dual problem of choosing the best among the rival models and obtaining the best estimates (in some statistical sense) of the parameters involved in the selected model.[2,3] The first stage, in which the precise mathematical relationship applicable to the system is identified, is known as the specification stage and the second one, in which precise estimates of the parameters are obtained, is known as the estimation stage. Both these stages constitute the important goals of modeling.

B. Modeling Strategy

In some cases, when an experimenter starts with an object of modeling either the whole system or a part of it, he may have some knowledge about the possible mechanism. In others, he may not know anything about the system. In most practical cases, normally he may have a partial knowledge about the system.

If he does not have any knowledge about the system, he may resort to purely empirical modeling, while if he has complete knowledge about the system he may directly proceed with the estimation stage. In most cases, however, he may be inbetween these two broad aspects, which leads to the so-called mechanistic modeling. The modeling strategy may be summarized in Figure 1. The various modeling strategies are further discussed below.

1. Empirical Models

Though one would like to find the "real model" applicable to a physical situation, it may be invariably an elusive goal. Under these conditions, one is left with two choices, viz., empirical modeling and mechanistic modeling. These empirical models may be polynomials in the independent variables. A typical empirical model may be represented by:[4]

$$y_i = \theta_o + \theta_1 \xi_{1i} + \theta_2 \xi_{2i} \ldots + \theta_p \xi_{pi} + \theta_{11} \xi_{1i}^2 + \theta_{22} \xi_{2i}^2 \ldots \quad (4)$$
$$+ \theta_{pp} \xi_{pi}^2 + \epsilon_1$$

where ϵ is the error associated with a measurement, θ and ξ refer to the corresponding parameters and independent variables respectively, and the subscript i refers to the i-th measurement.

Empirical models are arbitrarily chosen based on the apparent functional relationship of the response to the independent variables. Generally, either polynomials of the type given by Equation 4 or power-law models given by Equation 2 are used to represent empirical models. These have no relation, whatsoever, to the true mechanism of the system under consideration. Whenever the phenomenon under consideration is very complex, empirical models give useful guidance in predicting the response within the range of experimentation.

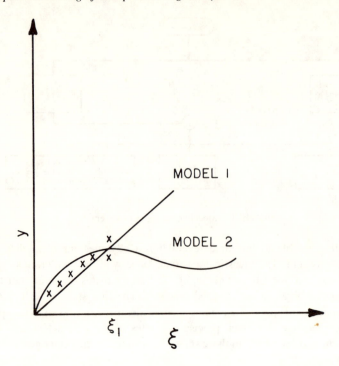

FIGURE 2. Experimental observations of a hypothetical situation — a plot of y vs. ξ.

2. Mechanistic Models

A mechanistic model is a mathematical relationship between the response (dependent variable) and the independent variables derived from a consideration of a plausible mechanism. For example, consider a simple heterogeneously catalyzed surface reaction

$$A.s + B.s \rightarrow C.s + s \tag{5}$$

where A.s, B.s, and C.s represent the adsorbed species A, B, and C respectively and "s" denotes a vacant site.

If surface reaction is rate-controlling, the model may be represented by

$$y = \frac{\theta_1 \xi_1 \xi_2}{(1 + \theta_2 \xi_1 + \theta_3 \xi_2)^2} \tag{6}$$

where y is the rate of reaction (dependent variable), and ξ_1 and ξ_2 are the gas-phase concentrations of the species A and B respectively.

Mechanistic modeling should be used judiciously. It is justified when the state-of-the-art is sufficiently advanced to make a useful mechanistic model. Judgment is needed in deciding when and when not to use mechanistic models.[1]

Two important aspects of mechanistic modeling consist of parameter estimation and model discrimination. Consider a set of observations, y, and the corresponding values of an independent variable ξ available on a particular system. The experimental observations of a hypothetical situation are shown on the y vs. ξ plot in Figure 2. Also shown on the graph are two different models arising out of different mechanistic considerations. From the graph, it is evident that either of Models 1 and 2 describe the system under consideration. It is seen from the graph that as long as the experimental data are obtained in the region $0 \leqslant \xi$

$\leq \xi_1$, it is difficult to say which of the two models governs the system. If the goal were to discriminate among these two models, the observations should be taken at higher values of the independent variable, ξ. Thus the "design of experiments" plays a dominant role in the model discrimination problem and its importance has been recognized long since.[3-14] Parameter estimation, model discrimination, and design of experiments for either model discrimination or parameter estimation constitute the important components of modeling. These topics are reviewed in the following sections.

II. ESTIMATION OF PARAMETERS IN LINEAR MODELS

A. Linear and Nonlinear Models
A model can be either linear or nonlinear. If the partial derivative of the response function with respect to any parameter does not contain the parameters, it is known as a linear model. For example, the model

$$y = \theta_1 \xi_1^m + \theta_2 \xi_2^n + \epsilon \tag{7}$$

is a linear model in the parameters. On the other hand the model

$$y = \theta_1 \exp(-\theta_2 \xi_1 \xi_2) + \epsilon \tag{8}$$

is a nonlinear model.

The technique to be employed for the estimation of parameters in any model depends on the nature of the model and its error structure. The parameter estimation techniques are reviewed briefly in the following sections. For the details of the methods one is referred to References 1 to 3 and 15 to 17.

B. Linear Regression
1. Estimation of Parameters
It is assumed that there are n measurements, y_1, made on a system where the true value of the quantity which is measured by y_i is called η_i and may be perfectly represented by:

$$\eta_i = \theta_1 \xi_{i1} + \theta_2 \xi_{i2} + \ldots + \theta_p \xi_{ip}, \qquad p \leq n \tag{9}$$

The quantities ξ_j, $j = 1, 2 \ldots p$ may be considered to be known perfectly. Furthermore, each measurement y_i may be considered the sum of the true value η_i for the ξ_{ij} which prevails and an error G_i with zero expectation. For example:

$$y_i = \theta_1 \xi_{i1} + \theta_2 \xi_{i2} + \ldots + \theta_p \xi_{ip} + \xi_i, \qquad i = 1, 2, \ldots n \tag{10}$$

where ξ_{ij} is the value of the j-th independent variable corresponding to the i-th measurement.

As a further assumption, it is assumed that the errors are uniformly and independently distributed. In other words, the covariance matrix of the errors may be given by:

$$V(\underline{\xi}) = I_{(nxn)}\sigma^2 \tag{11}$$

where I is an (nxn) identity matrix and σ^2 is the unknown variance.

The least squares estimates of the parameters ($\hat{\theta}$) are those values of the parameters which minimize the sum of squares of the differences between the observations and the predictions. These estimators of the parameters, also known as BLUE (Best Linear Unbiased Estimators) can be shown to be given by

$$\underline{\hat{\theta}} = (\underline{\xi}'\underline{\xi})^{-1} (\underline{\xi}'\underline{y}) \tag{12}$$

where $\underline{\xi}$ is given by:

$$\underline{\xi} = \begin{bmatrix} \xi_{11} & \xi_{12} & \cdots & \cdots & \xi_{1p} \\ \xi_{21} & \xi_{22} & \cdots & \cdots & \xi_{2p} \\ \cdot & \cdot & \cdots & \cdots & \cdot \\ \cdot & \cdot & \cdots & \cdots & \cdot \\ \xi_{n1} & \xi_{n2} & \cdots & \cdots & \xi_{np} \end{bmatrix} \tag{13}$$

$$\underline{y} = (y_1 y_2 \ldots\ldots y_n)' \tag{14}$$

2. Confidence Intervals and Confidence Regions

For a linear model of the type given by Equation 10, the $(1-\alpha/2)$ 100% confidence interval for the parameters θ_i is given by[1,2,15]

$$\theta_1 \pm t_{\nu, \alpha/2}[v(\theta_i)]^{1/2}$$

where $t_{\nu,\alpha/2}$ is the $(1 - \alpha/2)$ 100% point of the t-distribution with ν degrees of freedom. If the errors in the response are assumed to be normally and independently distributed with constant variance σ^2, then $v(\theta_i)$ is the i-th diagonal element of the matrix $(\underline{\xi}'\underline{\xi})^{-1}\sigma^2$. If the parameter estimates are correlated, then the joint $(1 - \alpha)$100% confidence region is given by:[1,2,15]

$$(\underline{\theta} - \underline{\hat{\theta}})'\underline{\xi}'\underline{\xi}(\underline{\theta}-\underline{\hat{\theta}}) = s^2 \, p \, F_\alpha \, (p,\nu) \tag{15}$$

where $\underline{\theta}$ is a vector of parameter values, which is being estimated under linear regression theory by $\underline{\hat{\theta}}$, s^2 is an independent estimate of experimental error variance σ^2, p is the number of parameters, and F_α (p,ν) is the critical value of F at a significance level of α.

III. ESTIMATION OF PARAMETERS IN NONLINEAR MODELS

Invariably, in several practical situations, one encounters nonlinear models. For example, for the gaseous dehydration of ethanol on a resin catalyst, Kabel and Johanson[18] gave the rate expression as:

$$r = \frac{k \, K_a(P_a^2 - P_e P_w/K_{eq})}{(1 + K_a P_a + K_w P_w + K_e P_e)^2} \tag{16}$$

In this expression, r is the measured (dependent) variable; P_a, P_e, and P_w are the controllable quantities (independent variables); and k, K_a, K_e, K_w, and K_{eq} are the parameters to be estimated.

For scientific reasons, the parameters in this nonlinear model should be estimated using nonlinear regression. In estimating the parameters in a linear model, one sets the partial derivatives of the residual sum of squares (RSS) with respect to each parameter to zero and solves the resulting equations simultaneously. However, in the case of nonlinear models, such equations cannot be easily solved. An iterative solution has been suggested[15,19] for estimating the parameters in a nonlinear model. The procedure consists of expanding the model by a Taylor's series expansion, retaining the linear terms, and then solving for the parameters by linear regression analysis. By retaining the linear terms in the Taylor's series expansion, the model may be written by

$$\eta_u = \eta_{uo} + \sum_{i=1}^{P} \left. \frac{\partial f(\underline{\xi}_u, \underline{\theta})}{\partial \theta_i} \right|_{\underline{\theta} = \underline{\theta}^o} (\theta_i - \theta_i^o) \tag{17}$$

Including experimental error, Equation 17 may be written as

$$z_u = \sum_{i=1}^{P} f'_{iu} \delta_i + \epsilon_u \tag{18}$$

where

$$z_u = y_u - f(\underline{\xi}_u, \underline{\theta}) \Big|_{\underline{\theta} = \underline{\theta}_o} + \epsilon \tag{19}$$

$$f'_{iu} = \left. \frac{\partial f(\underline{\xi}_u, \underline{\theta})}{\partial \theta_i} \right|_{\underline{\theta} = \underline{\theta}^o} \tag{20}$$

and

$$\delta_i = \theta_i - \theta_i^o \tag{21}$$

Since Equation 18 is linear, the correction vector can be obtained by applying linear regression.

The improved parameter estimates for the next trial are given by

$$\theta_i^{(1)} = \theta_i^o + \delta_i \tag{22}$$

This procedure is iterated, until the corrections δ_i become exceedingly small.

One can also apply the steepest descent procedures to determine parameter values.[3] This consists of setting up a first order design in the parameter space about a set of parameter estimates, calculating the direction of steepest descent, obtaining the minimum and using a second order design for the precise location of the minimum. This approach has been found to converge for nearly any set of initial parameter estimates, but its convergence can be agonizingly slow.

Marquardt[20] has suggested a compromise between these two primary methods which finds extensive application in estimating parameters in a nonlinear model.

It is very important to note that a proper choice of the initial parameter estimates is important in the estimation procedure. Intrinsically nonlinear models may be linearized to obtain initial estimates of the parameters. In some situations, one might have a prior knowledge of the probable values of the parameters which may be used as initial estimates.

Sequential simplex method has been successfully used for estimating the parameters in nonlinear models. This method is opportunistic and it converges even when the initial simplex straddles two or more ridges.[21-22]

A derivative free algorithm (called DUD) has been developed by Ralston and Jennrich for fitting models defined by systems of nonlinear differential equations. On a number of test problems, DUD has been claimed to compete favorably with even the best derivative-based algorithms.[24]

IV. MODEL DISCRIMINATION — SINGLE RESPONSE MODELS

The problem of discriminating among a set of competing models is that of choosing which one, hopefully unique, gives predictions which in some sense are better than those given by the others. Model discrimination is achieved by one of the following methods:

1.   Use of nonintrinsic parameters
2.   Use of intrinsic parameters
3.   Use of Bayesian methods
4.   Use of non-Bayesian methods

Several reviews[2,4,17,25] have appeared on the model discrimination aspect of which the salient features are discussed briefly in the following sections.

A. Use of Nonintrinsic Parameters

A nonintrinsic parameter is one which is introduced into a model for the purpose of discrimination from a set of rival models. Use of nonintrinsic parameters for model discrimination has been applied by Mezaki and Kittrell[26] for discriminating between two models based on single site and dual site mechanisms for modeling the vapor-phase rate of dehydration of secondary butyl alcohol to the olefin over a commercial cracking catalyst.

Consider the discrimination between two rival models A and B given by

$$\eta_A = f(\underline{\theta}, \underline{\xi})$$

and

$$\eta_B = g(\underline{\phi}, \underline{\xi}) \tag{23}$$

An equation of the type

$$y - 1/2 (\eta_A + \eta_B) = \lambda (\eta_B - \eta_A) \tag{24}$$

is regressed for λ. If λ were $-1/2$ then A is the correct model and if it were $1/2$, model B is the correct one. In practice λ may not be exactly equal to either $+1/2$ or $-1/2$. In such cases the confidence limits of λ should be assessed to see whether it includes $+1/2$ or $-1/2$ or both.

A somewhat analogous method consists of regressing the equation

$$y = (1 - \lambda) \eta_A + \lambda\eta_B \tag{25}$$

for λ. If λ were equal to 1, then B is the correct model. On the other hand, if λ were equal to zero, then A is the correct model. As before, it is recommended that the value of λ should be considered in its confidence range.

B. Use of Intrinsic Parameters

An intrinsic parameter is one which is inherently present in the model. These parameters, which are of a simpler functional form than the entire model, facilitate the experimenter's ability to test the adequacy of a proposed model. Kittrell and Mezaki[2,27] used this approach for proposing a model for the olefinic dehydration of a pure alcohol feed to a reactor. For details of this method one is referred to Kittrell and Mezaki's original article.[27]

C. Use of Bayesian Methods

The Bayes' theorem provides a useful means of discriminating among rival models. Bayes' theorem states.[28]

$$P(A_i/B) = \frac{P(A_i)\ P(B/A_i)}{\sum\limits_{i=1}^{r} P(B/A_i)\ P(A_i)} \tag{26}$$

where: A_i (i = 1 ... r) denotes the i-th model, B denotes the data, $P(A_i)$ denotes the prior probability of the i-th model, and $P(B/A_i)$ denotes the likelihood for the i-th model.

The Bayes' theorem requires knowledge of the prior probabilities of the various models. If one does not have knowledge of the prior probabilities of the various models, he can assign equal probabilities to all models.

The calculation of likelihood requires a knowledge of the error structure. If one assumes that the errors are normally and independently distributed with a variance σ^2, the probability density function (p.d.f.) of n observations $y_1, y_2....y_n$ is given by:[14]

$$p.d.f. = \frac{1}{(\sqrt{2\pi}\sigma^n} \exp - \frac{1}{2\sigma^2} \sum_{u=1}^{n} (y_u - f(\underline{\theta}, \underline{\xi}_u))^2 \tag{27}$$

On the other hand, if one assumes that the errors are normally distributed, but correlated with a variance-covariance matrix $\underline{V}\sigma^2$, then the p.d.f. of n observations $y_1, y_2...y_n$ is given by:

$$p.d.f. = \frac{1}{(\sqrt{2\pi}\sigma)^n \sqrt{|v|}} \exp - \frac{1}{2\sigma^2} [(\underline{y} - f(\underline{\theta}, \underline{\xi}))'\ \underline{V}^{-1}\ (\underline{y} - f(\underline{\theta}, \underline{\xi}))] \tag{28}$$

Once the observations ($\underline{y}$) and the design matrix ($\underline{\xi}$) are substituted into one of the above expressions for the p.d.f., the resulting expression, which is a function of $\underline{\theta}$ and σ^2, is called the likelihood.

Example of Model Discrimination — Reilly[4] considered a simple numerical example of discriminating among rival models. Assuming $\underline{\xi}$ as a single independent variable and y_u as the dependent variable, the set of numerical data shown in Table 1, was used for discriminating among rival models.

The following three different models were considered by Reilly for discrimination purposes.

$$\begin{array}{ll} \text{Model 1:} & y_u = \theta_{11}\ \xi_u + \epsilon_u \\ \text{Model 2:} & y_u = \theta_{21} + \theta_{22}\ \xi_u + \epsilon_u \\ \text{Model 3:} & y_u = \theta_{31} \exp \theta_{32}\ \xi_u + \epsilon_u \end{array} \tag{29}$$

The errors were assumed to be normally and independently distributed. The probability density function, defined by Equation 27, is applicable for this case. Using the least squares criterion, the parameters were estimated and the maximum likelihoods were calculated. Assuming equal prior probabilities, the posterior probabilities were calculated for each model as shown in Table 2. From the posterior probabilities it is seen that Model 3 is preferred.

This simple example illustrates how the Bayesian approach may be used in model discrimination.

Several investigators[1,13,29,30] have used Bayesian methods in model discrimination successfully.

Table 1
NUMERICAL DATA FOR
MODEL FITTING

u	$\underline{\xi}_u$	y_u
1	0	−1.290
2	1	5.318
3	2	7.049
4	3	19.886

Table 2
POSTERIOR PROBABILITIES OF
MODELS

Model	Prior probability	Likelihood	Posterior probability
1	0.3333	0.05	0.0002
2	0.3333	1	0.0049
3	0.3333	202.2	0.9948

D. Likelihood Discrimination

On quasi-religious grounds some people object to the use of Bayesian methods for the presentation of scientific results.[4,22,23] Likelihood discrimination shows the flexibility of Bayesian methods combined with an ability to "let the data speak for themselves".

To illustrate this approach, assume the i-th model to be represented by the functional relationship

$$\eta_{iu} = f_i(\underline{\theta}, \underline{\xi}_u) \tag{30}$$

Let $\underline{y}$ denote the vector of observations. Also assume that the observations are correlated with a variance-covariance matrix $\underline{V}\sigma^2$. Assume that n observations are available. Under these assumptions, the probability density function is given by Equation 28.

The maximum likelihood function (maximum w.r.t. σ^2) for the i-th model can be shown to be given by:[23]

$$L_i(\underline{\theta}, \sigma^2) = \frac{n^{n/2}}{(2\pi)^{n/2} M_i^{n/2} \sqrt{|\underline{V}|}} \exp(-n/2) \tag{31}$$

where M_i is the weighted sum of squares $(\underline{y} - \underline{\eta}_i)'\underline{V}^{-1}(\underline{y} - \underline{\eta}_i)$.

The ratio of maximum likelihoods for two models i and j, also known as the likelihood ratio, is a comparison of how well the two models can be made to fit the data.

For the above situation, then, one can show that the maximum likelihood ratio is given by:

$$\lambda = (L_i)_{max}/(L_j)_{max} = (M_j/M_i)^{n/2} \tag{32}$$

This ratio, λ, denotes the likelihood odds of model i versus model j.

The likelihood ratio method of discriminating among several rival models comprises finding the likelihood ratio between the best model (having minimum weighted sum of squares) and the other models taken one at a time. Thus, two-way comparisons are made by examining these ratios. This method (using the ratio of the maximum likelihoods) is not the only way of discriminating nuisance parameters in likelihood inference.

Rao et al.[23] and Reilly[4] have clearly demonstrated the use of maximum likelihood ratio as a statistical criterion for discriminating among rival kinetic models. Even with a moderate number of experimental observations, the likelihood ratio appeared to be a powerful one for discriminating among a large number of competing models.

There are inherent difficulties with any discrimination method when the models have different numbers of parameters.

V. TESTS OF MODEL ADEQUACY

Once the most plausible model has been obtained from a set of competing models, it is necessary to test for the adequacy of the model. This is usually accomplished by lack-of-fit F-tests and also by a residual analysis.

A. Lack-of-Fit F Tests

The F-statistic in model fitting may be defined by:

$$\frac{\text{lack of fit mean square}}{\text{pure error mean square}}$$

This ratio is compared to the critical value of F at the required degree of confidence and the corresponding degrees of freedom. If the above ratio is greater than F_{crit}, then the model is inadequate. For a detailed treatment on lack-of-fit F tests, one is referred to Reference 1 or 15.

B. Residual Analysis

A residual is defined as the difference between the observed and the predicted values of a response (i.e., $y - \eta$). The residual analysis can be applied either to the whole model as it is, or in some other cases (like in diagnostic checkup in model building) on some estimated parameters of a tentatively entertained theoretical model, so as to pinpoint the inadequacies.

As an illustration, assume that the true model of a set of observations is given by:

$$y_1 = \theta_o + \theta_1 \xi_1 + \theta_2 \xi_2 + \epsilon \tag{33}$$

Assume that the fit is given by:

$$\hat{y}_1 = \hat{\theta}_o + \hat{\theta}_1 \xi_1 \tag{34}$$

It will be seen that the residual $y_1 - \hat{y}_1$ is correlated with the variable ξ_2 in a linear way. Thus, if the residual is correlated with respect to ξ_2, it is indicative that the model should include a term consisting of ξ_2. Once this linear term is included and a residual analysis is performed, then the residuals should be random if the model were adequate. Time plots of residuals, as well as plots of residuals with respect to various controllable variables can detect possible model inadequacies, which can throw light on how to improve the model. This is the principle in adaptive model building. Kittrell et al.[31] have further demonstrated the use of diagnostic parameters for kinetic model building.

For a more detailed treatment on residual analysis one is referred to References 2 and 15.

VI. DESIGN OF EXPERIMENTS FOR MODEL DISCRIMINATION — SINGLE RESPONSE MODELS

In the past, for kinetic modeling, the "one-factor-at-a-time" method,[32] in which the experimental factors are varied one at a time, with the remaining factors held constant, has been used. However, this method of experimentation is found to be ineffective especially when there is interaction among the various factors under consideration.

When discrimination cannot be achieved from the existing data, it becomes necessary for the experimenter to design further experiments that provide maximum discrimination. It might be expected that all one has to do is to plan the future experiments carefully, run them and then it may be expected that one of the models would emerge as the best one. But this is not the case in general. Different researchers claim different models for the same phenomenon. A classic example is the water-gas shift reaction for which several models have been proposed. The reason for the discrepancy is that the "models are not put in jeopardy".

In the past, investigators took the difference in response given by the models for choosing a measure of discrimination.

One of the very early methods of design of experiments for discriminating between two rival models is that of Hunter and Reiner[6] who suggested maximizing the following design criterion:

$$D = (\hat{f}_{N+1}^N - \hat{g}_{N+1}^N) \tag{35}$$

where f_{N+1} and g_{N+1} denote the predicted responses for the two rival models for the $(N+1)$-th experimental trial (to be conducted) using the best estimates of parameters obtained after conducting N experimental trials.

When the number of rival models is greater than 2, Roth[13] suggests a criterion that involves choosing the experimental points that maximize the product of absolute differences in the predicted values of the response. The Roth criterion is given by:

$$C_{ji} = \prod_{\substack{k=1 \\ k \neq 1}}^{m} \eta_j^{(k)} - n_j^{(i)} \tag{36}$$

The subscript j corresponds to the values of the independent variables ξ_j. The design criterion to be maximized is the weighted average of the spread among response surfaces for the proposed models, weighted according to their probabilities,

$$D = \Sigma_i P_i^{(n)} C_{jy} \tag{37}$$

Both Hunter and Reiner's method, as well as Roth's method ignore the uncertainties associated with each model. To alleviate this difficulty, Box and Hill[7] proposed an excellent method for discriminating between m $(m \geq 2)$ rival models. Their method takes account of not only the difference in response given by the models to be discriminated among, but also the variance of the estimated response. The design criterion to be maximized is given by Box and Hill as

$$D = \frac{1}{2} \sum_{i=1}^{m} \sum_{j=1}^{m} \pi_{in} \pi_{jn} \left\{ \frac{(\sigma_i^2 - \sigma_j^2)}{(\sigma^2 + \sigma_i^2)(\sigma^2 + \sigma_j^2)} + \right.$$
$$\left. (\hat{y}_{n+1}^{(i)} - y_{n+1}^{(j)})^2 \left[\frac{1}{(\sigma^2 + \sigma_i^2)} \frac{1}{(\sigma^2 + \sigma_j^2)} \right] \right\} \tag{38}$$

Here σ^2 is the common variance of the n observations $y_1, y_2, \ldots y_n$, σ_i^2 and σ_j^2 are the variance for the predicted values of y_{n+1} under the models i and j respectively, and $\hat{y}_{n+1}^{(i)}$ and $y_{n+1}^{(j)}$ are the predicted values of y_{n+1} under models i and j respectively. Under certain assumptions, the p.d.f. was assumed to be given by:

$$p.d.f. = \frac{1}{\sqrt{2\pi(\sigma^2 + \sigma_i^2)}} \exp - \frac{1}{2(\sigma^2 + \sigma_i^2)} [(y_{n+1} - y_{n+1}^{(i)})^2] \tag{39}$$

The above criterion is obtained by maximizing an upper bound on the expected change in the entropy.

Hill and Hunter[33] extended the criterion to the case where σ^2 is not known. In the procedure suggested by Box and Hill, the design and analysis are carried out simultaneously and the stopping rule for experimentation is to stop when the posterior probabilities indicate that one model is clearly superior to the rest. Several studies[9,14,34] revealed that the model probabilities may oscillate considerably and Hill[35] suggests that the Box and Hill criterion must be applied very cautiously and the model should not be accepted too readily based on a small number of discriminating experiments.

Atkinson and Fedorov[36,37] advocated a criterion which is essentially a formalization of the intuitive idea of Hunter and Reiner.

Hsiang and Reilly[10] adopted a Bayesian procedure for model discrimination, which eliminated some of the stringent requirements of Box and Hill's procedure. However, it is claimed that their method requires excessive computer storage space for problems involving many parameters.

Atkinson and Cox,[38] and Atkinson and Fedorov[36,37] have developed the so-called "equal interest criterion" and the "T-optimality criterion" for the design of experiments for model discrimination.

The existing methods of experimental designs for discriminating between rival regression models have been reviewed recently by Hill.[35]

VII. DESIGN OF EXPERIMENTS FOR ESTIMATION OF PARAMETERS — SINGLE RESPONSE MODELS

Suppose that after suitable model-building experiments have been carried out, a given nonlinear model has been singled out as being adequate. Also assume that the space of the experimental variables is limited to some particular region of experimentation. Suppose it is desired to obtain estimates of the parameters $\underline{\theta}$ in this known model, given by Equation 3.

Let f_{iu} be the partial derivative of the nonlinear model with respect to any parameter θ_i evaluated at the u-th set of experimental conditions and taken at some set of parameter values $\underline{\theta}_o$, i.e.,

$$f_{iu} = \partial f(\underline{\theta}, \underline{\xi}_u)/\partial \theta_i \Big|_{\underline{\theta} = \underline{\theta}_o} \tag{40}$$

and p columns (parameters), may be written

$$\underline{F} = \{f_{iu}\} \tag{41}$$

Box and Lucas[11] have indicated that, under certain plausible assumptions, a choice of experimental points which will maximize $|\underline{F}'\underline{F}|$ will also be that choice of data points which will minimize the volume of the joint confidence region of parameters. $\sqrt{|\underline{F}'\underline{F}|}$ is inversely proportional to the volume of the joint confidence region.

Kittrell et al.[39] applied the above method for obtaining the precise estimates of parameters in the model for the reduction of nitric oxide, data for which were obtained by Ayen and Peters[40] and found that for the same number of data points, the parameters in the model can be estimated 18 times more precisely than by another commonly used, one-factor-at-a-time design.

Certain aspects of sequential design procedures for precise parameter estimation were discussed by Hosten and Emig.[41]

VIII. PARAMETER ESTIMATION — MULTIRESPONSE MODELS

In some cases, for a given set of experimental conditions, not one but a number of responses can be measured in a process. There are numerous examples of such systems where several responses are obtained. The use of multiresponse techniques increases the precision and accuracy of the parameter estimates and decreases the volume of the joint confidence ellipsoid. Singh and Rao[42] reviewed parameter estimation and model discrimination in multiresponse models in a recent review.

A. Mathematical Formulation

A multiresponse model can be denoted by:

$$\underline{Y} = \underline{\eta} \, (\underline{\theta}, \underline{\xi}) + \underline{\epsilon} \tag{42}$$

where η represents the true value of the response y, ϵ corresponds to the error associated with the measurement, $\underline{\theta}$ is a (px1) vector of parameters, and $\underline{\xi}_u$ is an (sx1) vector of controllable variables. More explicitly, the i-th response ($1 \leq i \leq r$) for the u-th experiment ($1 \leq u \leq n$) may be denoted by

$$y_u^{(i)} = \eta_u^{(i)} \, (\underline{\theta}, \underline{\xi}_u) + \epsilon_u^{(i)} \tag{43}$$

It is assumed that the errors are such that:

$$E \, (\epsilon_u^{(i)}) = 0 \text{ for all i, u}$$
$$E \, (\epsilon_u^{(i)} \, \epsilon_v^{(j)}) = 0 \text{ for all i, j, } u \neq v$$
$$E \, (\epsilon_u^{(i)} \, \epsilon_u^{(j)}) = \sigma_{ij} \text{ for i, j, u} \tag{44}$$

The vector $\underline{y}_u = (y^{(1)}{}_u, y_u^{(2)} \ldots \ldots y_u^{(r)})'$ of r responses for the u-th experiment has a symmetric covariance matrix Σ given by:

$$\Sigma = \{\sigma_{ij}\} \qquad \begin{aligned} & i = 1, 2 \ldots . r \\ & j = 1, 2 \ldots . r \end{aligned} \tag{45}$$

Just as before, the problem before us is to first discriminate among rival models of the type given by Equation 42 and second to estimate the parameters once the most adequate model has been obtained.

B. Parameter Estimation

In a situation of r responses and p parameters which are common to all the r response equations, Box and Draper[43] showed that the point minimization of

$$z = \sum_{i=1}^{r} \sum_{j=1}^{r} \sigma^{ij} v_{ij} \tag{46}$$

Here σ^{ij} is the (i,j)-th element of Σ^{-1} and gives the generalization of the method of least squares.

$$v_{ij} = \sum_{u=1}^{n} (y_u^{(i)} - \eta_u^{(i)}) (y_u^{(j)} - \eta_u^{(j)}) \tag{47}$$

When σ_{ij} is unknown, they suggest the minimization of

$$z_o = \sum_{i=1}^{r} \sum_{j=1}^{r} v_{ij} V_{ij}/r \tag{48}$$

for the estimation of parameters, where V_{ij} is the cofactor of v_{ij} in $|v_{ij}|$. They, however, warn that the overall criterion is likely to be offset by the lack of fit in a particular response. In such situations, it is safe to check for the adequacy of fit by a residual analysis and also by a consideration of the consistency of information from various responses by comparing the posterior distributions.

Beauchamp and Cornell[44] suggested an iterative procedure for the estimation of parameters in a multiresponse system. They suggest the minimization of

$$\phi(\underline{\theta}) = [\underline{Y} - \underline{\eta}(\underline{\theta}, \underline{\xi})]'\Omega^{-1}[\underline{Y} - \underline{\eta}(\underline{\theta}, \underline{\xi})] \tag{49}$$

Here Ω is defined by

$$\Omega = E(\underline{\epsilon}\,\underline{\epsilon}') = \Sigma \otimes I, \tag{50}$$

where $\otimes$ is the Kronecker product.

As a starting point, the r response questions are expanded about a trial vector of parameters $\underline{\theta}^{\circ}$ to give:

$$\underline{\eta}(\underline{\theta}, \underline{\xi}) = \underline{\eta}(\underline{\theta}^{\circ}, \underline{\xi}) + X\,\underline{\delta}^{\circ} \tag{51}$$

where

$$x_i = \{x_{u1}^{(i)} \qquad u = 1, 2 \dots n \tag{52}$$
$$= 1, 2 \dots p$$

for

$$x_{y1}^{(i)} = \delta\eta^{(i)}(\underline{\theta}, \underline{\xi}_u)/\delta\theta_i \tag{53}$$

$$X = (x_1, x_2, \dots, x_r)' \tag{54}$$

and

$$\underline{\delta}^{\circ} = \underline{\theta} - \underline{\theta}^{\circ} \tag{55}$$

Modified Gauss-Newton method discussed by Hartley[45] can be used to estimate the least squares estimates of $\underline{\theta}$.

Hunter[46] proposes the following criteria for various situations:

(i) If σ^{ij} are known, the quantity $\sum_{i=1}^{r} \sum_{j=1}^{r} \sigma^{ij} v_{ij}$ should be minimized

(ii) If $\sigma_{ij} = 0$ for $i \neq j$, $\sum_{i=1}^{r} \sigma^{ii} \sum_{u=1}^{n} (y_u^{(i)} - \eta_u^{(i)})^2$ should be minimized

(iii) For cases where the variances are assumed equal or when certain responses can be measured more precisely than others, then

$$\sum_{i=1}^{r} \sum_{u=1}^{n} (y_u^{(i)})^2 \text{ should be minimized}$$

In all the above situations, it is assumed that σ^{ij} are known. For the case when σ^{ij} are not known, minimization of Box and Draper's determinant criterion[43] gives the best estimates of $\underline{\theta}$ in a multiresponse model. Their determinant criterion is given by:

$$D_c = \left| \begin{array}{cccccc} \sum_{u=1}^{n} (y_u^{(1)} - \eta_u^{(1)})^2 & \cdots\cdots & \sum_{u=1}^{n} (y_u^{(1)} - \eta_u^{(1)}) (y_u^{(r)} - \eta_u^{(r)}) \\ - & - & - & - & - & - \\ - & - & - & \sum_{u=1}^{n} (y_u^{(r)} - \eta_u^{(u)})^2 & - \end{array} \right| \qquad (56)$$

Mezaki and Butt[47] applied the above criterion to a complex reaction sequence and found the criterion to provide an effective means of estimation of parameters. As regards convergence to final estimates, they observed that the determinant criterion to be much more rapid than the generalized nonlinear least squares techniques. They found very little difference in these two procedures so far as computational effort is concerned.

From a precision point of view, the determinant criterion puts the greatest weight on those responses which are measured most accurately, while the least squares criterion places the greatest weight on the data known less precisely.

Erjavec[48] points out that the error variance of each of the responses must remain constant from run to run.

Box et al.[49] discussed the possibility of one or the other kind of linear relationships which might exist among the responses. They suggest that if m linear relationships are known to exist then m dependent responses must be deleted before analyzing the data.

Box et al.[50] proposed a method which minimizes $|v_{ij}|$ with respect to $\underline{\theta}$ and Y_m for handling missing data $\underline{Y}_m$.*

IX. MODEL DISCRIMINATION — MULTIRESPONSE MODELS

Just like for single response models, the task before an experimenter is to discriminate among various competing multiresponse models. Either the likelihood discrimination techniques described earlier for the single response models or the Bayesian methods which make use of posterior probabilities can be used for model discrimination.

The likelihood $L_{n,k}$ for the k-th multiresponse model is given by:[5,30]

$$L_{n,k} (\underline{\theta}_k \, \Sigma_{n,k} \, , \, \underline{y}) = \frac{|\Sigma_{n,k}|^{1/2}}{(2\pi)^{r/2}} \exp \left[-\tfrac{1}{2} (\underline{y} - \underline{\eta}_{n,k})' \, \Sigma_{n+k}^{-1} \, (\underline{y} - \underline{\eta}_{n,k}) \right] \qquad (57)$$

where

$$\Sigma_{n,k} = \Sigma + X_{n,k} \, M_k^{-1} \, X_{n,k}' \qquad (58)$$

Here $X_{n,k}$ is an (rxp) matrix of partial derivatives whose (i,1)-th element is the partial derivative of the i-th response with respect to θ_1. M_k itself is given by:

$$M_k = \sum_{i=1}^{r} \sum_{j=1}^{r} \sigma^{ij} \, X_k^{(i)'} \, X_k^{(j)} \qquad (59)$$

---

* Reilly and Patinp-Leal[62] presented results of a study of the functional case of the problem of parameter estimation when there is error in all the variables. Their study leads to new and efficient algorithms for finding point estimates and their precisions.

Likelihood discrimination is carried out by comparing maximum likelihoods for different models, taken one at a time.

In Bayesian methods, discrimination is assumed to be achieved if one of the models attains a high posterior probability. At the end of n experiments, if it is not possible to discriminate among the various competing models, further experimentation is necessary which is conducted by a sequential design of experiments as described below.

In Bayesian methods, in general, maximum likelihood is used for computing the posterior model probabilities. However, the likelihood, being a function of the parameters, which are themselves random variables, is in itself a random variable. Prasad and Rao[30] used expected likelihood in place of point or maximum likelihood in computing the posterior probabilities and demonstrated the utility of expected likelihood in efficient model discrimination.

X. DESIGN OF EXPERIMENTS FOR MODEL DISCRIMINATION — MULTIRESPONSE MODELS

Hill and Hunter[51] extended Box and Hill's criterion for application to multiresponse models. Their discrimination criterion, to be maximized for obtaining the experimental conditions for the (n + 1)-th experimental run, is given by:

$$
\begin{aligned}
D = \tfrac{1}{2} \sum_{h=1}^{m} \sum_{k=h+1}^{m} P_{n,h}\, P_{n,k}\, \{ \text{trace } [\Sigma_{n+1,h}\, \Sigma_{n+1,k}^{-1} + \Sigma_{n+1,k}\, \Sigma_{n+1,h}^{-1} - 2\, I_r] \\
+ (\hat{\underline{Y}}_{n+1,h} - \hat{\underline{Y}}_{n+1,k})'\, (\Sigma_{n+1,h}^{-1} + \Sigma_{n+1,k}^{-1})\, (\hat{\underline{Y}}_{n+1,h} - \hat{\underline{Y}}_{-n+1,k}) \}
\end{aligned}
\tag{60}
$$

where I_r is an (rxr) identity matrix, $\hat{\underline{Y}}_{n+1,k}$ is an (rx1) vector of response for the n + 1-th experimental conditions predicted using the k-th model and the previous best estimates of parameters, and $P_{n,k}$ is the posterior probability of model k after the observations $\underline{Y}_n$ are obtained.

The design of experiments and analysis is conducted sequentially until discrimination is achieved. Because of the presence of prior probabilities in B, less emphasis is placed on poorly fitting models.

An important point to be noted about the Box-Hill procedure is about the stopping rule. Box-Hill criterion suggests that discrimination is achieved when the posterior probability of one of the models is rather high compared to those of other models. However, several investigators[9,14,34,52] experience show that model probabilities in certain situations may oscillate considerably from stage to stage. The rule must be applied cautiously and a model should not be accepted readily on the basis of a small number of discriminating experiments. Despite the limitations of Box-Hill's procedure, it has been successfully applied in various situations.

Another useful criterion for the design of experiments for model discrimination has been proposed by Roth.[13] For a multiresponse system, Roth's criterion is given by:

$$
D = \prod_{i=1}^{r} \left(\sum_{k=1}^{r} (P_{n,k} \prod_{\substack{h=1 \\ h \neq k}}^{m}) (\hat{Y}_{n+1,h}^{(i)} - \hat{Y}_{n+1,k}^{(i)}) \right)
\tag{61}
$$

where $P_{n,k}$, $\hat{Y}_{n+1,h}^{(i)}$ are defined earlier.

Roth's criterion consists of maximizing the divergence between the values of responses predicted by different models. If the parameters are not known to the same degree of accuracy in all the models, a misleading set of experimental conditions may be predicted.

Reilly and Blau[53] observed that in the use of Roth's criterion the large divergences due to inaccurately known parameters outweigh the smaller divergences due to accurately known parameters, resulting in a probable wrong sequential design.

XI. DESIGN OF EXPERIMENTS FOR PARAMETER ESTIMATION — MULTIRESPONSE MODELS

Once the precise model applicable to a particular physicochemical situation has been determined, the goal of the experimenter is to estimate the parameters of the model precisely, which needs additional experimentation. Draper and Hunter[54] proposed a method to conduct n* additional experiments. Their approach consists of maximizing the posterior density of $\underline{\theta}$, after conducting (n + n*) runs, with respect to $\underline{\theta}$ and n* — which is equivalent to maximizing the determinant

$$D = \sum_{i=1}^{r} \sum_{j=1}^{r} \sigma^{ij} \underline{X}_i' \, \underline{X}_j \qquad (62)$$

where $\underline{X}_i$ is an (n + n*) x p matrix of elements defined by Equation 52 the derivatives being evaluated at the current best estimates of $\underline{\theta}$.

Box[55,56] discussed the problem of nonlinear model building in situations when constancy of covariance matrix cannot be assumed. He has also developed a computer program which undertakes the task of formal computations of optimal experimental designs.

In the program, he proposed the use of numerical differentiation for obtaining the derivatives of the model responses with respect to the unknown parameters. The program can be extended to handle errors in the input variables.

Box[57] has also derived an experimental design for estimating only q parameters of interest out of a total p, parameters. For a locally uniform prior distribution for $\underline{\theta}$, his criterion consists of maximizing the determinant

$$D = |A_{11} - A_{12} \, A_{22}^{-1} \, A_{12}'| \qquad (63)$$

where A_{11} is a (qxq) matrix, and A_{12}, A_{22} are obtained by partitioning $\underline{A}$, given by

$$A = \sum_{u=1}^{n} \sum_{i=1}^{r} \sum_{j=1}^{r} \sigma^{ij} \, X_{iu} \, X_{ju}' \qquad (64)$$

for

$$X_{iu} = \left\{ \frac{\partial \eta^{(i)} (\hat{\theta}, \underline{\xi}_u)}{\partial \, \theta_1} \cdots \cdots \frac{\partial \eta^{(i)}(\hat{\theta}, \underline{\xi}_u)}{\partial \, \theta_p} \right\} \qquad (65)$$

all derivatives being evaluated at $\hat{\underline{\theta}}$.

XII. A CASE STUDY ON SEQUENTIAL MODEL DISCRIMINATION IN MULTIRESPONSE SYSTEMS

Prasad and Rao[30] advocated the use of expected likelihood in efficient model discrimination. From the basic definition, expected likelihood is given by:

$$E \, (L_{n+1,u} \, (\underline{\theta}|\Sigma, \, \underline{Y}_{n+1})) = \int_{-\infty}^{\infty} \cdots \int_{-\infty}^{\infty} L_{n+1,u} \, (\underline{\theta}|\Sigma, \, \underline{Y}_{n+1})$$
$$P_{n,u} \, (\underline{\theta}) \, d\theta_1 \, d\theta_2 \, \ldots \, d\theta_p \qquad (66)$$

One is referred to Prasad and Rao's article for details on evaluating the integral in Equation 66.

Either the point likelihood (Equation 57) or the expected likelihood (Equation 66) can be used to update model probabilities using the Bayes' theorem. Prasad and Rao presented computational results using multiresponse data to demonstrate the utility of the expected likelihood in efficient model discrimination in a catalyst fouling system. Two discriminatory criteria, viz., the Box-Hill criterion (Equation 60) and Roth's criterion (Equation 61) are compared.

The extensive multiresponse experimental data on the product distribution for various space velocities and decay times for the vapor phase reaction between tetrachloroethane and a large excess of chlorine on activated silica gel, collected by Prasad and Doraiswamy,[58] are used in the present case study on model discrimination. The reaction proceeded according to the following scheme.

$$C_2H_2Cl_4 \xrightarrow{Cl_2} C_2HCl_5 \xrightarrow{Cl_2} C_2Cl_6$$
$$(A_1) \qquad\qquad (A_2) \qquad\qquad (A_3)$$

Their experiments revealed a decline in the catalyst activity with time. Typical data used for illustrative purposes are given in Table 3. The prior information consisted of a subset of 6 experimental data points, while the remaining 64 data points were used for sequential design purposes.

Under certain assumptions a general reaction rate model for the system under consideration may be represented by two simultaneous equations of the type:

$$\eta = f_1 (\underline{\xi}, \underline{\theta}) \, f_2(a) \tag{67}$$

$$da/dt = f_3 (\underline{\xi}, \underline{\theta}) \, f_4(a) \tag{68}$$

where η is the rate of reaction, f_1, f_2, f_3, and f_4 are functions depending on the system, t is the decay time, and a is the activity of the catalyst ($a = \eta_{t=t}/\eta_{t=0}$).

For the chlorination reaction, 11 different combinations of fouling reactions were considered by Prasad and Rao resulting in 11 different plausible models.

f_1 Consists of two functions corresponding to the two independent responses (η_1), the rate of disappearance of A_1, and (η_2), the rate of formation of A_2. For example:

$$\eta_1 = - \, \theta_1 \xi_1 \, a \tag{69}$$

and

$$\eta_2 = (\theta_1 \xi_1 - \theta_2 \xi_2) \, a \tag{70}$$

A further assumption is given by:

$$f_2(a) = f_4(a) = a \tag{71}$$

The different combinations of fouling reactions considered by the authors and the corresponding model equations are given in Table 4.

The complex method of Box[59] was used for estimating the parameters in various models.

The prior information consisted of six random experiments (experiments numbered 8, 13, 18, 23, 41, and 46). Since no one model was preferred to start with, prior probability of each model was assumed to be 1/11. The following initial estimates were used for parameter estimation:

Table 3

TYPICAL EXPERIMENTAL RATE DATA[29]

S. no.	Exp. no.	Space time (g hr/mol)	Mol % of A_1 (100 ξ_1)	Mol % of A_1 (100 ξ_2)	Decay time (hr) (ξ_3)	y_1	y_2
						mol/g hr $\times$ 10^2	
1	13	90	44.3	52.0	1	−0.460	0.242
2	58	30	82.1	17.4	8	−0.515	0.609
3	36	15	89.0	10.9	5	−0.715	0.668
4	46	60	63.2	35.2	6	−0.518	0.450
—	—	—	—	—	—	—	—
—	—	—	—	—	—	—	—
70	64	15	90.3	9.5	9	−0.585	0.524

$$\theta_1 = 0.01 \text{ hr}^{-1}, \; \theta_2 = 0.001 \text{ hr}^{-1}, \; \theta_3 = 0.03 \text{ hr}^{-1}$$

The sequential discrimination procedure consisted of the following steps:

1.   From the prior information, the parameters of each model were estimated.
2.   Using the current best parameter values and either the Box-Hill criterion (Equation 60) or Roth's criterion (Equation 61) the next best discriminating experiment was designed.
3.   Using the additional data of step 2 after conducting the experiment, the parameters of all the models were updated.
4.   The probabilities of all the models were updated using either the point likelihood or the expected likelihood in the Bayes' theorem.

If the posterior probability of any one model is exceedingly high, discrimination is achieved. Otherwise, the above sequence of steps were repeated.

Typical values of the posterior probabilities of the best model obtained by the above procedure are given in Table 5.

The following salient features were observed by this case study:

1.   An entirely different set of discriminating experiments was designed with the same prior information depending on the discriminatory criterion used.
2.   The Box-Hill criterion proved to be more efficient in comparison to Roth's criterion.
3.   Convergence towards the best model was faster with the expected likelihood than with the point likelihood.

Caution should be exercised with either the point likelihood method or the expected likelihood method. Sometimes, the model probabilities oscillate from run to run. The superiority of any discriminatory method depends both on the models as well as the data.

XIII. EVOLUTIONARY OPERATIONS AND RESPONSE SURFACE METHODOLOGY

In most industrial operations, the aim is to improve the performance of the process without interrupting the normal operation of the plant. In the Evolutionary Operation (EVOP) technique, the response is measured at certain levels of the variables in a repeated way such that the uncertainty associated with changing the levels of the variables is smaller than the effect of this change. Once better settings of the variables are discovered in this way, they can be used to improve the performance of the plant.

Table 4
RIVAL MODELS[29]

Model no.	Fouling reactions	Model equations
1	$A_1 \rightarrow P$	$\eta^1 = -\theta_1\xi_1 \exp(-\theta_3\xi_1\xi_3)$ $\eta^2 = (\theta_1\xi_1 - \theta_2\xi_2)\exp(-\theta_3\xi_1\xi_3)$
2	$A_2 \rightarrow P$	$\eta^1 = -\theta_1\xi_1 \exp(-\theta_3\xi_2\xi_3)$ $\eta^2 = (\theta_1\xi_1 - \theta_2\xi_2)\exp(-\theta_3\xi_2\xi_3)$
3	$A_3 \rightarrow P$	$\eta^1 = -\theta_1\xi_1 \exp(-\theta_3(1 - \xi_1 - \xi_2)\xi_3)$ $\eta^2 = (\theta_1\xi_1 - \theta_2\xi_2)\exp(-\theta_3(1 - \xi_1 - \xi_2)\xi_3)$
4	$A_1 + A_2 \rightarrow P$	$\eta^1 = -\theta_1\xi_1 \exp(-\theta_3\xi_1\xi_2\xi_3)$ $\eta^2 = (\theta_1\xi_1 - \theta_2\xi_2)\exp(-\theta_3\xi_1\xi_2\xi_3)$
5	$A_1 + A_3 \rightarrow P$	$\eta^1 = \theta_1\xi_1 \exp[-\theta_3\xi_1(1 - \xi_1 - \xi_2)\xi_3]$ $\eta^2 = (\theta_1\xi_1 - \theta_2\xi_2)\exp[-\theta_3\xi_1(1 - \xi_1 - \xi_2)\xi_3]$
6	$A_2 + A_3 \rightarrow P$	$\eta^1 = -\theta_1\xi_1 \exp[-\theta_3\xi_2(1 - \xi_1 - \xi_2)\xi_3]$ $\eta^2 = (\theta_1\xi_1 - \theta_2\xi_2)\exp[-\theta_3\xi_2(1 - \xi_1 - \xi_2)\xi_3]$
7	$A_1 + A_2 + A_3 \rightarrow P$	$\eta^1 = -\theta_1\xi_1 \exp[-\theta_3\xi_1\xi_2(1 - \xi_1 - \xi_2)\xi_3]$ $\eta^2 = (\theta_1\xi_1 - \theta_2\xi_2)\exp[-\theta_3\xi_1\xi_2(1 - \xi_1 - \xi_2)\xi_3]$
8	$A_1 \rightarrow P$ $A_2 \rightarrow P$	$\eta^1 = -\theta_1\xi_1 \exp[-\theta_3(\xi_1 + \xi_2)\xi_3]$ $\eta^2 = (\theta_1\xi_1 - \theta_2\xi_2)\exp[-\theta_3(\xi_1 + \xi_2)\xi_3]$
9	$A_1 \rightarrow P$ $A_3 \rightarrow P$	$\eta^1 = -\theta_1\xi_1 \exp[-\theta_3(1 - \xi_2)\xi_3]$ $\eta^2 = (\theta_1\xi_1 - \theta_2\xi_2)\exp[-\theta_3(1 - \xi_2)\xi_3]$
10	$A_2 \rightarrow P$ $A_3 \rightarrow P$	$\eta^1 = -\theta_1\xi_1 \exp[-\theta_3(1 - \xi_1)\xi_3]$ $\eta^2 = (\theta_1\xi_1 - \theta_2\xi_2)\exp[-\theta_3(1 - \xi_1)\xi_3]$
11	$A_1 \rightarrow P$ $A_2 \rightarrow P$ $A_3 \rightarrow P$	$\eta^1 = -\theta_1\xi_1 \exp(-\theta_3\xi_3)$ $\eta^2 = (\theta_1\xi_1 - \theta_2\xi_2)\exp(-\theta_3\xi_3)$

Table 5
POSTERIOR PROBABILITIES OF BEST MODEL USING THE BOX-HILL CRITERION AND EITHER POINT LIKELIHOOD OR EXPECTED LIKELIHOOD

Discrimination stage	Total number of experiments	Posterior probability of model 9 using	
		Point likelihood $\pi_{n+1,9}$	Expected likelihood $\pi_{n+1,9}$
1	7	0.502	0.506
2	8	0.513	0.635
3	9	0.623	0.735
4	10	0.656	0.814
5	11	0.730	0.842

Another, closely resembling method of improving the performance of the process plant is via response surface methodology. Here the response surface is represented by an empirical relationship in the independent variables. The principle applied makes use of the fact that the response can be represented by a linear relationship in the independent variables, if it is far from the optimum. In the vicinity of optimum, a linear representation of the response surface is inadequate and one has to use better (e.g., quadratic) representation of the response surface, which can lead to proper identification of optimal conditions for plant operation.

For detailed treatment on EVOP and response surface methodology, one is referred to the original articles by Hunter and Kittrell;[60] and Kittrell and Erjavec,[61] respectively.

XIV. CONCLUDING REMARKS

Parameter estimation, model discrimination, and experimental design constitute the important components of modeling. Various computer-based techniques have been developed in the recent past for parameter estimation. Linear models are easy to handle, while nonlinear models require initial estimates of parameters in the estimation problem. If the models are nonlinear, but are intrinsically linear, then they can be linearized to obtain the initial estimates after which one of the several nonlinear estimation techniques can be applied for estimating the parameters. In recent years, derivative-free methods have been developed which can possibly save the computation time.

In model discrimination, both the Bayesian as well as the likelihood methods find extensive application. Also, the sequential design techniques find extensive use both for model discrimination as well as for parameter estimation. Experimental designs for simultaneous model discrimination and parameter estimation need to be tried in the future.

Parameter estimation techniques are well developed for multiresponse situations.

While the various modeling techniques have been applied successfully so far on single response systems, their application to multiresponse systems needs to be verified on "real systems".

ACKNOWLEDGMENTS

One of the authors (MSR) acknowledges a visiting professorship offered by LSU, Baton Rouge, LA during which period this manuscript was prepared.

NOMENCLATURE

a	activity
A_i	i-th model; also used for the determinant defined by Equation 64
A_1, A_2, A_3	reacting species
B	data
C	concentration as in Equation 2; also used for spread among response surfaces as in Equation 36
D	design criterion
D_c	determinant criterion as given in Equation 56
E	expected likelihood as defined by Equation 66
f	functional relationship
F	matrix of partial derivatives as defined by Equation 41; also used for F in F-distribution
g	functional relationship
I_r	(rxr) identity matrix
k	rate constant
K_{eq}	equilibrium constant
K	adsorption equilibrium constant
L	likelihood
M_k	as defined by Equation 59
n	number of experimental observations
p	number of parameters; also used for partial pressure as in Equation 16
P	probability
$P(A_i)$	prior probability of the i-th model

$P(B/A_i)$	likelihood for the i-th model
r	rate of reaction
s	vacant site on a catalyst surface
s^2	estimate of variance
t	time coordinate; also used for t in t-distribution
v	covariance
V	variance-covariance matrix
V_{ij}	cofactor of v_{ij} in v_{ij}
X	partial derivative matrix as defined by Equation #52
y	observed value of response
z	residual as defined by Equation 19; also used for the quantity defined by Equation 46
z_0	quantity defined by Equation 48

Greek Symbols

Ω	$E(\underline{\epsilon}\,\underline{\epsilon}')$
α	significance level
δ	quantity defined by Equation 21
ϵ	experimental error
η	true or predicted value of response
Σ	covariance matrix as defined by Equation 45
$\underline{\theta}$	vector of parameters
$\underline{\phi}$	vector of parameters
$\underline{\xi}$	vector of input variables
σ^2	variance
σ	element of variance-covariance matrix (if subscripts are used) or inverse of variance-covariance matrix (if superscripts are used)
ϕ	functional relationship
π	3.1426; also used for probability
λ	nonintrinsic parameter; also used for likelihood ratio as in Equation 32
$\otimes$	Kronecker product

Subscripts

a,b	species A and B
A,B,C,E,W	species A, B, C, E, and W
o	initial value
i,j	response or model
r	number of responses

Superscripts

i,k	models
m,n	reaction orders
o	initial value
^	best estimate
'	transpose

REFERENCES

1. **Box, G. E. P., Hunter, W. G., and Stuart, J. S.,** *Statistics for Experimenters,* John Wiley & Sons, New York, 1978, chap. 16.
2. **Kittrell, J. R.,** Mathematical modeling of chemical reactions, in *Advances in Chemical Engineering,* Vol. 8, Drew, T. B., Cokelet, G. R., Hoopes, J. W., Jr., and Vermeulen, T., Eds., Academic Press, New York, 1970, 97.
3. **Kittre., J. R., Mezaki, R., and Watson, C. C.,** Estimation of parameters for nonlinear least squares analysis, *Ind. Eng. Chem.,* 57(12), 19, 1965.
4. **Reilly, P. M.,** Statistical methods in model discrimination, *Can. J. Chem. Eng.,* 48, 168, 1970.
5. **Hill, W. J.,** Statistical Techniques for Model-Building, Ph.D. thesis, University of Wisconsin, Madison, 1966.
6. **Hunter, W. G. and Reiner, A. M.,** Designs for discriminating between two rival models, *Technometrics,* 7, 307, 1965.
7. **Box, G. E. P. and Hill, W. J.,** Discrimination among mechanistic models, *Technometrics,* 9, 57, 1967.
8. **Hunter, W. G. and Mezaki, R.,** An experimental design strategy for discriminating among rival mechanistic models, *Can. J. Chem. Eng.,* 45, 247, 1967.
9. **Froment, G. F. and Mezaki, R.,** Sequential discrimination and estimation procedures for rate modeling in heterogeneous catalysis, *Chem. Eng. Sci.,* 25, 293, 1970.
10. **Hsiang, T. and Reilly, P. M.,** A practical method for discriminating among mechanistic models, *Can. J. Chem. Eng.,* 49, 865, 1971.
11. **Box, G. E. P. and Lucas, H. L.,** Design of experiments in nonlinear situations, *Biometrika,* 46, 77, 1959.
12. **Hosten, L. H.,** Non Bayesian sequential experimental design procedures for optimal discrimination between rival models, *Proceedings 4th Int. Symp. on Chemical Reaction Eng.,* Heidelberg, 1976.
13. **Roth, P. M.,** Design of Experiments for Discrimination Among Rival Models, Ph.D. thesis, Princeton University, Princeton, 1965.
14. **Hill, P. D. H.,** Optimal Experimental Designs for Model Discrimination, Ph.D. thesis, University of Glasgow, Glasgow, 1976.
15. **Draper, N. R. and Smith, H.,** *Applied Regression Analysis,* John Wiley & Sons, New York, 1966.
16. **Plackett, R. L.,** *Principles of Regression Analysis,* Oxford University Press, Oxford, England, 1960.
17. **Froment, G. F.,** Model discrimination and parameter estimation in heterogeneous catalysis, *AIChE J.,* 21, 1041, 1975.
18. **Kabel, R. L. and Johanson, L. N.,** Reaction Kinetics and adsorption equilibria in the vapor-phase dehydration of ethanol, *AIChE J.,* 8, 621, 1962.
19. **Bard, Y.,** *Nonlinear Parameter Estimation,* Academic Press, New York, 1974.
20. **Marquardt, D. L.,** An algorithm for least squares estimation of non-linear parameters, *J. Soc. Ind. Appl. Math.,* 2, 431, 1963.
21. **Nelder, J. A. and Mead, R.,** A simplex method for function minimization, *Comput. J.,* 7, 308, 1965.
22. **Rao, M. S.,** Identification of Rate Controlling Steps in Pentene-2-Hydrogenation from Selectivity — A Maximum-Likelihood Ratio Approach, Ph.D. thesis, University of Waterloo, Ontario, Canada, 1969.
23. **Rao, M. S., Hudgins, R. R., Reilly, P. M., and Silveston, P. L.,** A selectivity method for modelling the kinetics of pentene-2 hydrogenation, *Can. J. Chem. Eng.,* 49, 354, 1971.
24. **Ralston, M. L. and Jennrich, R. I.,** Dud — a derivative-free algorithm for nonlinear least squares, *Technometrics,* 20, 7, 1978.
25. **Gaver, K. M. and Geisel, M. S.,** Discriminating among rival models: Bayesian and non-Bayesian methods, in *Frontiers of Econometrics,* Zarembka, P., Ed., Academic Press, New York, 1974, 49.
26. **Mezaki, R. and Kittrell, J. R.,** Discrimination between two rival models through non-intrinsic parameters, *Can. J. Chem. Eng.,* 44, 285, 1966.
27. **Kittrell, J. R. and Mezaki, R.,** Discrimination among rival Hougen-Watson models through intrinsic parameters, *AIChE J.,* 13, 389, 1967.
28. **Volk, W.,** *Applied Statistics for Chemical Engineers,* McGraw-Hill, New York, 1958.
29. **Prasad, K. B. S.,** Use of Expected Likelihood in Sequential Model Discrimination, M. Tech. thesis, Indian Institute of Technology, Kanpur, India, 1976.
30. **Prasad, K. B. S. and Rao, M. S.,** Use of expected likelihood in sequential model discrimination in multiresponse systems, *Chem. Eng. Sci.,* 32, 1411, 1977.
31. **Kittrell, J. R., Hunter, W. G., and Mezaki, R.,** The use of diagnostic parameters for kinetic model building, *AIChE J.,* 5, 1014, 1966.
32. **Yang, K. H. and Hougen, O. A.,** Determination of mechanism of catalyzed gaseous reactions, *Chem. Eng. Prog.,* 46, 146, 1950.
33. **Hill, W. J. and Hunter, W. G.,** A note on designs for model discrimination: variance unknown case, *Technometrics,* 11, 396, 1969.

34. **Kentzheimer, W. W.,** Modelling of Heterogeneous Catalysed Reactions using Statistical Experimental Design and Data Analysis, University of Pennsylvania, Philadelphia, 1970.
35. **Hill, P. D. H.,** A review of experimental design procedures for regression model discrimination, *Technometrics*, 20(1), 15, 1978.
36. **Atkinson, A. C. and Fedorov, V. V.,** The design of experiments for discriminating between two rival models, *Biometrika*, 62, 57, 1975.
37. **Atkinson, A. C. and Fedorov, V. V.,** Optimal design: experiments for discriminating between several models, *Biometrika*, 62, 289, 1975.
38. **Atkinson, A. C. and Cox, D. R.,** Planning experiments for discriminating between models, *J. R. Statist. Soc. B*, 36, 321, 1974.
39. **Kittrell, J. R., Hunter, W. G., and Watson, C. C.,** Obtaining precise parameter estimates for nonlinear catalytic rate models, *AIChE J.*, 12(1), 5, 1966.
40. **Ayen, R. J. and Peters, M. S.,** Catalytic reduction of nitric oxide, *Ind. Eng. Chem. Proc. Design Dev.*, 1, 204, 1962.
41. **Hosten, L. H. and Emig, G.,** Sequential experimental design procedures for precise parameter estimation in ordinary differential equations, *Chem. Eng. Sci.*, 30, 1357, 1975.
42. **Singh, S. and Rao, M. S.,** Parameter estimation and model discrimination in multiresponse models, *Ind. Chem. Eng.*, 23(2), 19, 1981.
43. **Box, G. E. P. and Draper, N. R.,** The Bayesian estimation of common parameters from several responses, *Biometrika*, 52, 355, 1965.
44. **Beauchamp, J. J. and Cornell, R. G.,** Simultaneous nonlinear estimation, *Technometrics*, 8, 319, 1966.
45. **Hartley, O.,** The modified Gauss-Newton method for the fitting of non-linear regression functions by least squares, *Technometrics*, 3, 269, 1961.
46. **Hunter, W. G.,** Estimation of unknown constants from multiresponse data, *Ind. Eng. Chem. Fund.*, 6, 461, 1967.
47. **Mezaki, R. and Butt, J. B.,** Estimation of rate constants from multiresponse kinetic data, *Ind. Eng. Chem. Fund.*, 7, 120, 1968.
48. **Erjavec, J.,** Strategy for estimation of rate constants from isothermal reaction data, *Ind. Eng. Chem. Fund.*, 9, 187, 1970.
49. **Box, G. E. P., Hunter, W. G., MacGregor, J. F., and Erjavec, J.,** Some problems associated with the analysis of multiresponse data, *Technometrics*, 15, 33, 1973.
50. **Box, M. J., Draper, N. R., and Hunter, W. G.,** Missing values in multiresponse nonlinear model fitting, *Technometrics*, 12, 613, 1970.
51. **Hill, W. J., Hunter, W. G.,** Tech. Rep. No. 65, Department of Statistics, University of Wisconsin, Madison, 1966.
52. **Siddik, S. M.,** Kullback-Lieber Information Function and the Sequential Selection of Experiments to Discriminate among Several Linear Models, Ph.D. thesis, Princeton University, Princeton, N.J., 1972.
53. **Reilly, P. M. and Balu, G. E.,** The use of statistical methods to build mathematical models of chemical reacting systems, *Can. J. Chem. Eng.*, 52, 289, 1974.
54. **Draper, N. R. and Hunter, W. G.,** Design of experiments for parameter estimation in multiresponse situations, *Biometrika*, 53, 525, 1966.
55. **Box, M. J.,** Improved parameter estimation, *Technometrics*, 12, 219, 1970.
56. **Box, M. J.,** Simplified experimental design, *Technometrics*, 13, 19, 1971.
57. **Box, M. J.,** An experimental design criterion for precise estimation of a subset of the parameters in a nonlinear model, *Biometrika*, 58, 149, 1971.
58. **Prasad, K. B. S. and Doraiswamy, L. K.,** Effect of fouling in a fixed-bed reactor for a complex reaction: test of proposed model and formulation of an optimal policy, *J. Catal.*, 32, 384, 1974.
59. **Box, M. J.,** A new method of contrained optimization and a comparison with other methods, *Comput. J.*, 8, 42, 1965.
60. **Hunter, W. G. and Kittrell, J. R.,** Evolutionary operation: a review, *Technometrics*, 8, 389, 1966.
61. **Kittrell, J. R. and Erjavec, J.,** Response surface methods in heterogeneous kinetic modeling, *Ind. Eng. Chem. Proc. Des. Dev.*, 7, 321, 1968.
62. **Reilly, P. M. and Patino-Leal, H.,** A Bayesian study of the error in variables model, *Technometrics*, 23(3), 221, 1981.
63. **Iyengor, S. and Rao, M. S.,** Statistical techniques in modeling of complex systems: single and multiresponse models, *IEEE Trans. Syst. Man Cybernet.*, 13(2), 175, 1983.

Chapter 4

MODELING CANCER PROBLEMS AND ITS PROGRESS

W. Düchting

TABLE OF CONTENTS

I. INTRODUCTION

Everybody knows that a cancer cell has lost the ability to control growth and division. The purpose of this chapter is to show how control theory and computer simulation can be applied to tumor growth.

The rapid parallel advances in the fields of biological sciences have given the impetus to the development of numerous mathematical models of cell growth.[1] Some of them deal with stochastic theory of cell proliferation[2] or follow the theory of branching processes.[3]

In addition to this line of research an attempt is made to explain cell renewal processes as closed loop negative feedback circuits and to interpret malignant neoplastic cell growth as unstable control loops.[4]

Therefore this chapter summarizes several control models of cell growth which are essential for research in modern cell kinetics.

II. BASIC IDEA

The first approach to modeling the cancer problem started with a simple single-loop control circuit (Figure 1) looking for the stability conditions of the closed loop.[4] With rather simplified assumptions, the criterion for the stability of the regulator gene was developed when disturbances (ionizing radiation, carcinogenic chemicals, oncogenic viruses) had caused irreversible changes in the control-mechanism of the DNA. Step by step this defect transformed normal cells into tumor cells and finally led to an increase of the tumor cells (Figure 1). So cancer diseases may be interpreted as cell renewal systems which have become structurally unstable in their closed-loop control circuit.

III. REGULATORY MODEL OF THE BLOOD-FORMING PROCESS

A. Motivation for Study

The blood-forming process is a rather complex biological system itself (Figure 2). Therefore, the following considerations have been reduced to the special case of the erythropoiesis,[5] that is, the production of red blood cells. The reasons for modeling only this subsystem are as follows:

1. There is a dynamic balance between the cell renewal and the cell loss in the steady state. Approximately 2×10^{11} new erythrocytes have to be produced daily. This calls for an interpretation by the automatic control systems theory.
2. At present this field offers the most data gained in experiments.[6]
3. Outstanding oscillations, e.g., of a chronic leukemia (Figure 3) have been observed[7] which may be closely connected with stability problems of the closed-loop control circuit of the blood cells.

B. Description of the Model

The complexity of erythropoiesis has been reduced to the simplified block diagram in Figure 4. In the multi-loop erythropoietic feedback model of Figure 4 the reference variable R represents the required tissue oxygen symbolized by an equivalent number of erythrocytes. If a deviation $E = R - C$ arises between the reference input signal R and the controlled variable C (momentary number of red cells) the hormone erythropoietin produced mainly in the kidney affects the determined stem-cell compartment in the bone marrow to feed the proliferation pool with a certain number (Y32) of the determined stem cells. The switching function operator (SE 2) symbolizes the switching activities of the genes. Step by step the determined stem cells pass different pools in which cell division and cell differentiation take place, then as reticulocytes or erythrocytes C, they enter the peripheral blood and after a mean life span of about 120 days they are finally removed.

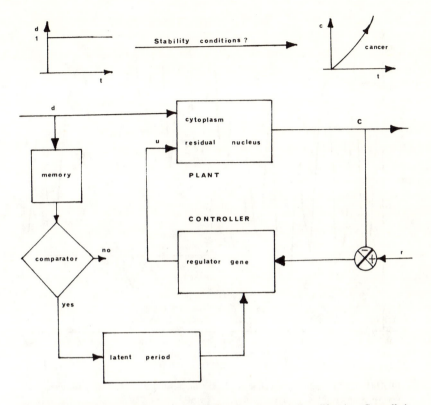

FIGURE 1. Block diagram of a control process of malignant cell proliferation. Controlled variable C: deviation of the number of cells from the steady state; Reference input r: e.g., hormones; Disturbances d: carcinogens, e.g., damps of tar, mechanical stimuli, viruses; Control signal u: production of specific regulation substances, e.g., enzymes.

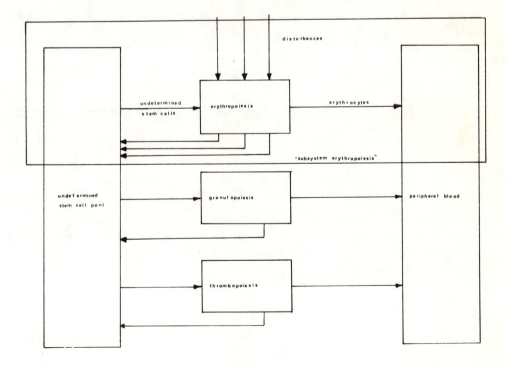

FIGURE 2. Blood-forming cell renewal system.

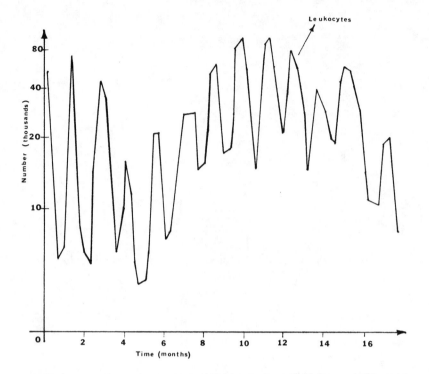

FIGURE 3. Oscillations of leukocytes in chronic myelogenous leukemia still occurring 1 year after onset of hydroxyurea therapy.[7]

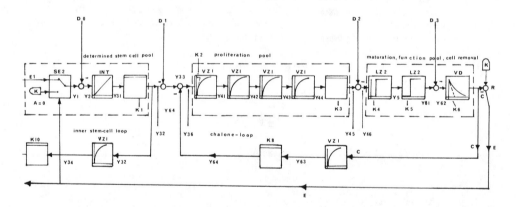

FIGURE 4. Multiloop erythropoietic feedback control block diagram. R: desired number of erythrocytes = required tissue oxygen = Reference input; E: Error = erythrocytes; D: Disturbance, e.g., viruses, raybleeding.

Interacting subsystems have led to two additional internal control loops. The chalone hypothesis with tissue-specific unspecified mitotic inhibitors[8] is represented by the right inner loop and the produced committed stem cells Y32, have a direct negative feedback on their own compartment along the left inner stem-cell loop.

C. Performance of Simulation

The time behavior of the single control-loop elements is described in Figure 4 by the respective time responses. Their notations were chosen corresponding to the terminology of the block-oriented programming language ASIM (Analogue SIMulation) of the firm AEG-Telefunken[9] which was used for the computer simulation. A listing of the corresponding computer program is shown in Figure 5. It contains the structure and output notations and the data statements of the parameters. The latter were chosen from the experimental data on cell production, cell division, their transportation, and life spans of red blood cells.[10] For the unknown parameter (K1) a feasible value was taken. The time constant (K2) and the gain (K3) alter with the produced quantity of erythropoietin. That is why the parameters are not constant in reality, but are interdependent and time-variant.

D. Simulation Results

For different cases the system has been stimulated by disturbances and subsequently the time response of the erythrocytes $C = f(T)$ has been determined by computer simulation.

A first example describes the aplastic anemia and results in a curve of erythrocyte response $C = f(T)$ shown in Figure 6. It represents the situation that at the time $T = 50$ days a sudden change of the parameter K10 of the inner stem-cell loop takes place. The increased feedback factor K10B leads to a conspicuously decreasing response of red blood cells (dashed line in Figure 6) which is similar to clinical observations of asplastic anemia. On the other hand, it has been shown that a sudden increase in the amplification factor of the inner loop from K8A = 0.02 to K8B = 0.09 at T = 50 days, leads to rising oscillations in the course of the red cell response (Figure 7) symbolizing certain forms of erythroleukemia or polycythemia.[11,12]

Thus, it is shown that the deterministic model developed for the blood forming process enables the study of totally different blood diseases.

IV. GENERALIZED MULTI-LOOP COMPARTMENT MODEL

A. Specifications to the Model

The erythropoiesis control-system configuration described in Figure 4 is unsuitable for studying the behavior of a single cell in the different compartments. Therefore, a generalized multi-loop and multi-compartmental model has been developed in Figure 8 which fulfills the following additional requirements:

- The existence and location of a single cell can be ascertained at any moment
- The mean life span of each cell can be variably fixed and controlled
- Each single cell can be eradicated at any desired moment by disturbances from outside

B. Description of the Model

According to the compartment hypothesis, the process of cell division and differentiation takes its course from undetermined stem cells in the bone marrow (symbolized by signal A03 of Figure 8) through various stages of development till they finally become mature cells. An example of these mature cells is the erythrocytes S3 which are sent into the peripheral blood.

Compartment 2 gets the impulses for forming new cells from the previous compartment 1. In this model special subroutines produce the input signals C21 when a cell is being divided and is advancing into the adjacent compartment. A digital logic device ascertains and registers the presence of each cell in the specific compartment. The total number of cells present in compartment 2 is symbolized by S2. S2 is compared in the control loop II with the desired number of cells S22, which represents for instance the required hormone value. If the controlled variable S2 sinks below the reference signal S22 by a certain fixed value the pulser SIØL sends an output signal S23 — symbolizing a humoral agent — to the stem-cell pool. This feedback signal is called SOO. The fictitious switch SE1 in the stem-cell pool, which symbolizes the gene activities, is temporarily closed and delivers a stem cell for the erythrocytic track.[13]

```
*-------------------REGELKREIS MIT TOTZEIT UND NICHTSTETIGEM ELEMENT
*-------------------STRUKTUR
Y1 = SE2 (XD,A,XD)
Y2 = INT(0,Y1)
Y2A,Y2B = SA(Z1,Y2)
Y3A = K1A * Y2A
Y3B = K1B * Y2B
Y3 = Y3A + Y3B
Y41 = VZ1(0,K2,Y31)
Y31 = Y3
Y62 = KB * Y61
Y61 = VZ1(0,K7,Y45)
Y42 = VZ1(0,K2,Y41)
Y43 = VZ1(0,K2,Y42)
Y44 = VZ1(0,K2,Y43)
Y44A, Y44B = SA(Z2,Y44)
Y45A = K3A * Y44A
Y45B = K3B * Y44B
Y45 = Y45A + Y45B
Y5 = LZ2(K4,Y45)
Y6 = LZ2(K5,Y5)
Y6A,Y6B = SA(Z3,Y6)
XA = VD(0,K6A,Y6A)
XB = VD(0,K6B,Y6B)
X = XA + XB
XD = W - X
A1 = 1
A2 = LZ1(TZ1,A1)
A3 = LZ1(TZ2,A1)
A4 = SIML(0,A2)
A5 = SIML(0,A3)
A6 = MEMR(0,A4,A5
*-------------------PARAMETER
A = 0
W = 100
H1 = ABS(XD) - 1.E10
K1A = 0.02
K1B = 0.025
K2 = 0.25
K3A = 12
K4 = 1
K5 = 2
K6A = 30
K6B = 10
K7 = 0.1
K8 = 0.2
Z1 = 0
Z2 = A6
Z3 = 0
```

FIGURE 5.   Listing of erythropoiesis control-system simulation (computer program ASIM).

```
TZ1 = 50
TZ2 = 0
K3B = 10
*-------------------BEARBEITUNG
SKIP H1
RZEIT (0.,0.1,160.)
PLOTTER (A4Q,T/D,15,X/ERYS,7)0.,160.,T, − 200.,200.,X
END
```

FIGURE 5. Continued.

C. Subsystem "Cell"

The basic element of each compartment (Figure 8) is an individual cell symbolized in Figure 9. Important fixed input signals are the dividing impulses C, the disturbance signal ZA, the mean life span TA, and the input channels UA, VA, NA to alter the life span of the single cell by external signals. The most important output signals are AA or AB which prove the existence of cells by the variables logic "1". The newly formed daughter cells symbolically enter the following compartments via the output RA.

The formulation of the conditions of the "artificial cell" in the ASIM-language can be taken from the block diagram in Figure 10.

D. Performance of Simulation

In this case again the block-oriented language ASIM is used for simulating the time behavior of the generalized compartment model.

The start impulses and the reference input signals (S11, S22, S33) are programmed in such a manner that the average number of cells is

- Two in compartment 1
- Four in compartment 2
- Eight in compartment 3

The mean value of the fictitious life span of cells among the single compartments (Figure 8) varies between 0.967 and 1.05 days and the standard deviation shows values between 0.197 and 0.333.[14]

E. Simulation Results

As the design of the generalized model is flexible, it is possible to carry out case studies by means of computer simulation which could not be performed in an in vitro or in an in vivo experiment.

If a hypothesis is made that all cells in compartment 1 are proliferating cells which are altogether destroyed at the time $T = 8.5$ days, then Figure 11 shows the time behavior of cells in the different compartments. It can be seen from this how the perturbation is transmitted in sequence to the adjacent compartments. It is also observed that immediately after the disturbance has taken place the inner control loop I (Figure 8) is activated by the feedback impulse S13. The loops II and III show a reaction, too, when after a time delay, the perturbation has reached the compartments 2 and 3. If a medical diagnosis based on the manifestation of the cell kinetics of compartment 3 were to be set up, it would be difficult to recognize the causes of the disease which, for example in Figure 11, are evoked by disturbances in compartment 1.

The most frequent case in reality is when single cells of different compartments, being in the proliferating phase, are destroyed directly by external or internal disturbances. A

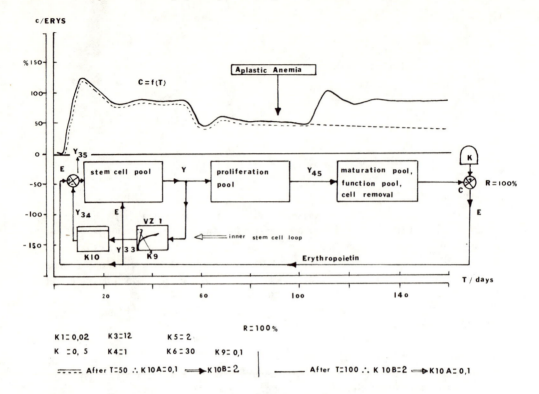

FIGURE 6.   Red cell responses C = f(T) to changes of the amplifier gain K10 of the inner stem-cell loop.

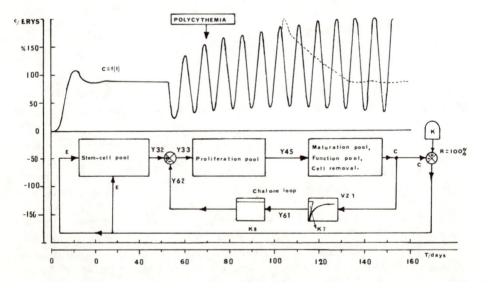

FIGURE 7.   Red blood responses C = f(T) to changes of the amplifier gain K8 of the inner chalone-loop.

random eradication of the cells Z1.1, Z2.3, and Z3.1 to Z3.5 is assumed at the time T = 8.5 days (Figure 12). At the same time, a halving of the life span of the cells in compartment 1 ensues as a *parametric* change. In addition, a *structural* alteration after T = 8.5 days causes the two additional impulses SZ1 and SZ2 to influence the feedback loop permanently. The simulation result is that the dynamics of the graphs outlined in Figure 12 show a surprising similarity with the time courses of an uncontrollable increase in the number of cells in the case of experimentally observed tumor progression.

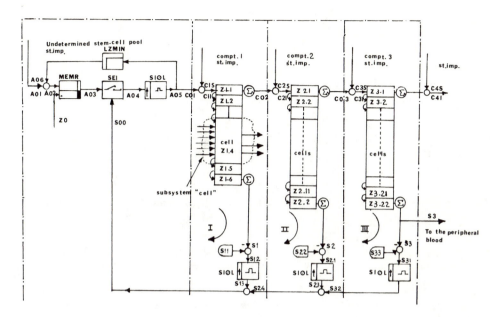

FIGURE 8. Simplified block diagram of a generalized cell-renewal feedback-control system. S1, S2, S3: number of cells in the appropriate compartment; S11, S22, S33: reference input signals; Σ: impulses for cell division; st. imp.: start impulses.

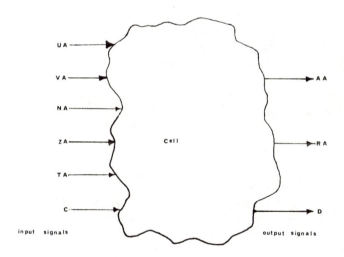

FIGURE 9. Subsystem "cell".

V. MODEL OF SELFREPRODUCING NORMAL AND MALIGNANT CELLS

A. Introduction

Most of the models developed so far to simulate the cell renewal systems have dealt with the determination of the time behavior of cell systems as a response to different perturbations (following the compartment hypothesis) or they considered cell kinetic problems of the different phases of the cell cycle.[15] They, however, disregarded the local configuration of cells and the cell-to-cell-interactions of adjacent cells in a tissue. This shortcoming is compensated in this section by an approach constructing a model which allows computer predictions about the structure of cell patterns and also about the cell kinetic dynamics of selfreproducing cell systems.

Basic questions concerning the structure of cell renewal systems with regard to information and automata theory have been answered in reports.[16-18]

B. Description of the Model

The starting point of constructing a model of a selfreproducing cell system[19] is a small homogeneous two-dimensional grid section of a tissue (Figure 13).

The model is to be organized in a way to satisfy the following conditions:

- To start the growing process of each single cell by an initial signal at any desired discrete moment
- To deliberately assign the mean life span of each cell to a life span matrix element (Figure 14) or to change it within a limited region
- To destroy each single cell at any desired discrete moment by external disturbances, e.g., by a surgical removal which can be simulated by a "reset signal" in the computer program
- To grow two competitive cell systems which differ in the life spans of their single cells, the origin of the cells being indicated by different markers in the corresponding transition matrix

Concerning the cellular communication between adjacent cells there are the following assumptions and rules of formation:

- A cell can become live only if there is at least one living neighbor cell in each row or column
- If a living cell is surrounded by dead or destroyed horizontal or vertical cells only, this cell is isolated and dies at once
- A cell division happens only if a dividing cell finds an empty space next to itself. Otherwise, no cell division takes place and the cell dies
- If there are several "vacancies" in the direct neighborhood of a cell, a pseudo-binary random generator decides which empty space is to be occupied by the daughter cell
- The cells in the border lines (Figure 13A) follow additional conditions. The state of a boundary cell is registered by all neighboring cells and is enclosed in the data processing mechanism, but division and reset signals are exchanged between this cell and its inner neighboring cell only

C. Performance of Simulation

From the viewpoint of automata theory, a selfreproducing cell system can be considered as an asynchronous state-controlled sequential circuit. In this model the outputs are not only dependent on the input signals, but also on the momentary interior states of the cells. Therefore, it is reasonable to study the dynamics of a selfreproducing cell system by means of a computer model of a state-controlled circuit, e.g., the microcomputer system 8080 of the firm INTEL®.

According to the specifications mentioned above, computer programs for a 10×10 matrix (Figure 14) have been set up. First, each cell of a 10×10 matrix is given a constant life span deliberately fixed according to an equipartition (Figure 14) with two cell systems (○ and ▲).

Then, the initial and reset instructions, the number of sequential runs, and the time step of the output are fed in. Controlled by the master program the cell matrix is treated, one element after the other, row by row. As a result, the local distribution of the living cells and their total number (without the border cells) are put out on a graphic display or a high-speed plotter after the pass time of $T = 1$. The next step is to reduce the life spans in the cell matrix (Figure 14) by the value 1 and to start the second run. When the life span of a cell gets to "zero", the cell is being divided provided that all boundary conditions regarding the adjacent cells are given.

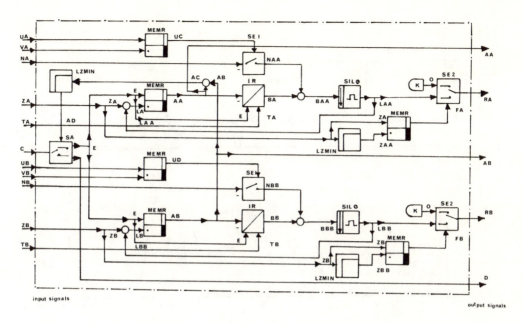

FIGURE 10.   Block diagram of a cell pair (subsystem).

D. Simulation Results

Assuming different initial and boundary conditions, it is possible to study the space and time behavior of disturbed selfreproducing cell systems by raising deliberated disturbances, e.g., reset instructions which are equivalent to a surgical removal.

Starting from three initial cells which at T = 1 time unit are rapidly growing in the center of the matrix it can be observed that the cell renewal process (Figure 15) following the given formation rules reaches the stationary state at T = 45 time units. When interpreting the curve Z = f(T), attention has to be paid to the fact that Z represents the number of living cells *without* the borderline cells. A disturbance at the time of T = 60 units killing all the cells of the rows 1 to 4 is already leveled at T = 75 time units.

Now, a completely occupied cell matrix at T = 1 time units is assumed (Figure 16) having a core of a 3 × 3 matrix with rapidly growing cells which could be interpreted as neoplastic tissue.

A surgical removal of the cell columns 1 to 7 at T = 101 time units shows that the only malignant cell which was not removed by this operation is able to increase rapidly to the value of about 40 cells after T = 3000 units. This is to demonstrate that a partial surgical removal of a tumor accelerates the proliferation of the only remaining tumor cell.

Finally, Figure 17 shows the change of the cell configuration if a neoplastic tissue is placed side by side to a normal one. After a long-term run of T = 3000 time units it can be seen that the malignant cells for the most part have infiltrated into the "normal" region and that neoplastic tissue is already dominant.

The simulation results show a great similarity to morphological cuts and cell kinetic curves both gained by experiments.

VI. MODELING THE SPREAD OF CANCER CELLS IN TISSUE

A. Introductory Remarks

In Section V the attempt of modeling spatial cell growth was restricted by the capacity of only 10 × 10 cells due to the application of a microcomputer.

Doing away with this disadvantage the present section develops a computer model for a

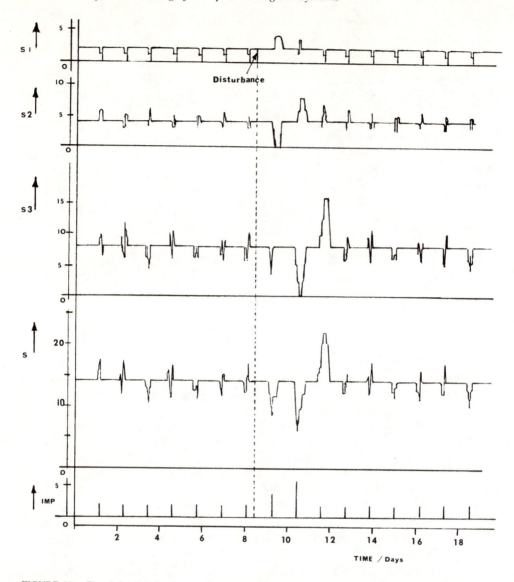

FIGURE 11.   Time behavior of cell renewal process with disturbances to all cells of compartment 1 at T = 8.5 days. S1, S2, S3: number of cells in the appropriate compartment. S: number of total cells (S = S1 + S2 + S3); IMP: feedback impulses.

cell space of a maximum of 100 × 100 cells. For this simulation a large digital computer (CYBER 72/76) is required. This step not only enables the simulation of four cell systems growing in competition, but also makes it possible to use algorithms which lead the divided tumor cells into the direction of the minimum number of cells of a column or a row and to take into consideration the loss of tumor cells. Thus, the attempt is made to get a model of malignant cell growth[20] more realistically similar to the medical and biological reality.

B. Specifications to the Model

In the following, specifications are given which enable the study of the cell renewal process in a two-dimensional 100 × 100 cell space. Up to three different cell systems characterized by Figures 1 to 3 shall be allowed to grow in competition. In addition, tumor cells characterized by the symbol (*) may come into existence and grow. A fictitious boundary symbolized by the special character (+) representing the organ boundary is introduced to limit cell growth of normally growing cell systems.

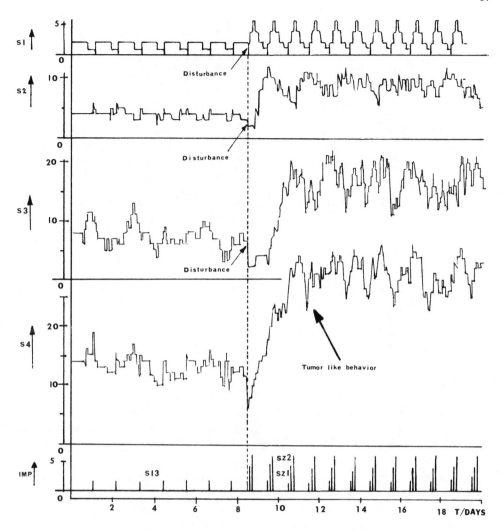

FIGURE 12. Time behavior of cell renewal process with disturbances to the cells Z1.1, Z2.3, and Z3.1 to Z3.5, halved mean life span of the cells in compartment 1; two additional regular impulses in the control loop I when T = 8.5 days.

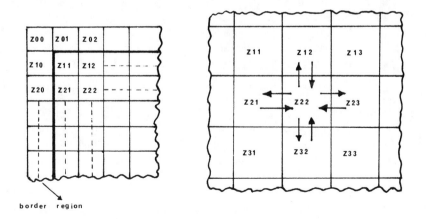

FIGURE 13. Grid configuration of cells.

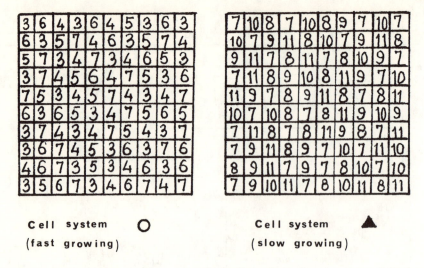

Cell system   O
(fast growing)

Cell system   ▲
(slow growing)

FIGURE 14.   Life span matrices of two different cell systems.

Growth patterns of normally growing cell systems may be generated by the following detailed rules which compared with those of Reference 19 have remarkably been extended and improved:

1. Each cell of a cell system can at any time be put in any position of the two-dimensional array.
2. The life span of a cell is determined by a pseudo-random number generator.
3. A cell can exchange information only with adjacent horizontal and vertical neighboring cells. This condition, as well as the restriction of the cell growth to a two-dimensional cell space serves to reduce the computing effort.
4. A division of a normal cell can take place only when there is an adjacent free position for one of the two new daughter cells. If there is none, the mother cell is deleted in the cell space when its normal life span is over.
5. If there are several empty places, a pseudo-random number generator determines the direction for one of the new-born daughter cells.
6. Each cell can at any time be deleted by an external reset signal, speaking in medical terms that could e.g., be a surgical removal or a radiation treatment.
7. The division of a normal cell will not take place when the new cell has to cross the prescribed internal border line.

In contrast to characteristics 1 to 7 applying to a *"normal cell"* the specifications 4 and 7 shall not be valid for a *"tumor cell"*. These are substituted and extended by the following notations:

8. The division of a tumor cell can take place even if there is no adjacent vacancy for one of the two daughter cells. In this case the dividing tumor cell finds out that direction which has a minimum of the total number of cells existing, and the division takes place toward this direction by shifting the cells of the corresponding row or column one position farther away. This assumption tries to take into account the influence of the tension of the neighboring cells on the dividing cell.
9. Resulting from assumption 8, a tumor cell can cross the prescribed internal border line of an organ and can grow into any other cell system. This behavior corresponds to the clinical observation of the cancellation of contact inhibition in the case of a malignant progression.

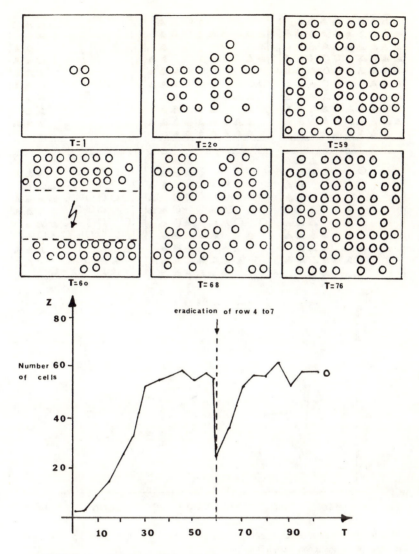

FIGURE 15.   Disturbed cell renewal process of the fast growing cell system (O).

10.   A pseudo-random number generator shall assign the life span to a tumor cell.
11.   The probability of tumor-cell loss has been fixed to be $p = 5\%$ at any time step. This specification represents a first step to simulate the cell loss of the tumor cells which in reality is by no means constant — as is supposed here, but dependent on the tumor volume, location, and time.

C. Performance of Simulation

At present, no complete analytic solution can be found when simulating the complex biological cell renewal system; it seems convenient to develop an algorithm each for the "cell renewal processes" and for the subsystem "cell" (Figure 18).

The simulation of the cell renewal process itself can be considered with the simplified aspect of a sequential row-to-row calculation of the cell algorithm for each single cell, the life span of a cell each time being reduced by one time unit after each finished run. The configuration gained in this way serves as the initial configuration of the following calculation step (Figure 18). This process is repeated, until the finish time of the simulation run has been arrived. When a cell has reached the given outside border line signified by the special sign ● there is an immediate interruption of the simulation run.

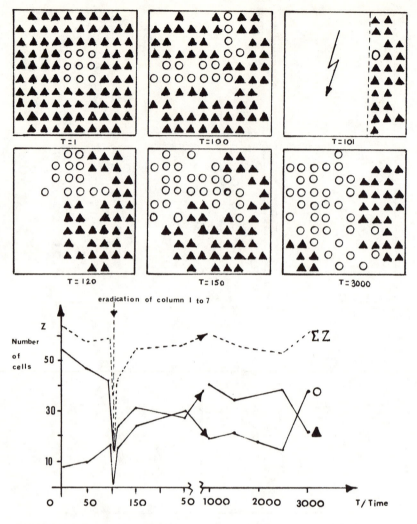

FIGURE 16.   Cell renewal process of two competing cell systems with a partial surgical removal of the malignant neoplastic tissue.

The formal input for a simulation run comprises:

- Details on the sequence of the systems to be occupied
- The nomination of the number of the data for each cell system
- The notation of the configuration data of the single systems including their border lines
- The initialization of the pseudo-random number generators
- The fixation of the final simulation time and the output time step
- The indication of the time data for the start of different events, e.g., the arising of tumor cells or their deletion

The results of a simulation run are, apart from the input data, the spatial structure of the cells in a two-dimensional cell space and the time course of the number of the cells. Furthermore, the layout of the program for special cases enables the coded output of important parameter values of the cells at any time (cell system, actual life span of a cell, cell generation, position of the cells).

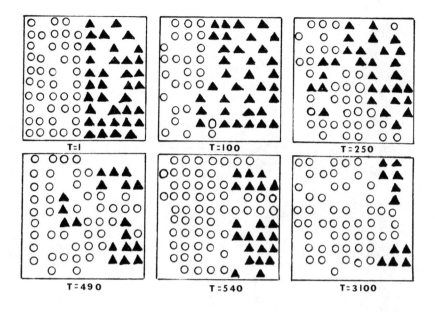

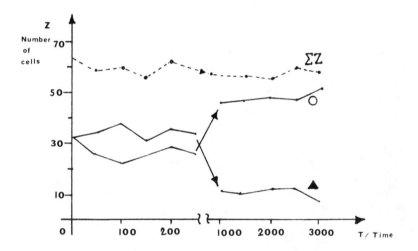

FIGURE 17. Cell renewal process of normal and neoplastic tissue touching each other.

D. Simulation Results

In principle, the simulation of organ forms whatsoever is possible. For reasons of distinctness, however, strongly simplified contours of the cell systems have been chosen for the input procedure of the individual simulation runs. In consequence of the given technical restrictions of the printer a quadratic output of a 100×100 cell array is not possible. There is a higher density of dots in the horizontal direction with all spatial cell space views of this section leading to a *nonquadratic* graph of the cell matrix, which must be taken into special account when interpreting the results.

The first case starts from three cell systems which are normally regenerated. The mean life spans of the cells are T1 = 10 ± 2; T2 = 20 ± 2; T3 = 30; ± 2 time units. If after T′ = 50 time units an inner partial area of cell system 1 is eliminated by a surgical removal — e.g., in a liver — it can be seen in Figure 20 that a normal regeneration of the disturbed system takes place without cell growth outside the border line of the system.

In contrast to this quite normal behavior, an internal boundary between the three cell

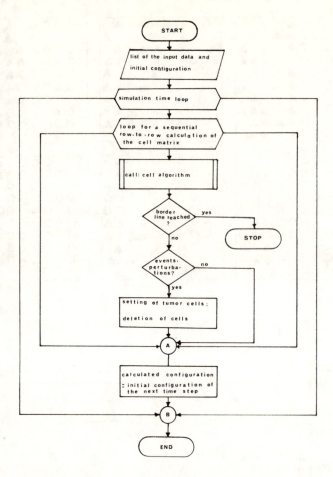

FIGURE 18.   Oversimplified flow chart of the cell renewal process.

systems is introduced in the next case. The spatial step-by-step increase of the assumed tumor nucleus shows the growth of the tumor cells beyond the organ borderlines into cell systems 2 and 3. Following the internal tension, these cell systems make their own cells move into the free external space. If the tumor is removed at $T' = 20$ time units, that means if most of the tumor cells are erased but not all of them, there is a rapid growth into the vacant area and a quick spatial expansion of the tumor.

Systematic computer simulation enables us to study the influence of the mean life span of a tumor cell (Figure 19) and of the size of a tumor nucleus (Figure 20) on tumor growth assuming constant cell loss of tumor cells.

A comparison of the tumor growth of cell systems with equal initial configuration but different life spans T^* of a tumor cell shows in Figure 19 that the very rapid tumor growth with $T^* = 5 \pm 2$ time units slows down with $T^* = 10 \pm 2$ time units and after $T = 120$ with $T^* = 15 \pm 2$ time units leads to a remission of the tumor. It follows that the mean life span of a tumor cell T^* is a decisive variable.

If, on the other hand, the mean life span of the tumor cells of $T^* = 15 \pm 2$ time units is maintained, but the initial configuration is reduced from an 11×11 tumor-cell matrix (Figure 20) to a 3×3 tumor-cell core the tumor cells die away.

It follows that a critical initial number of tumor cells of a tumor nucleus is necessary for the growth of a tumor. However, this is only a necessary criterion for tumor growth, not a sufficient one, because there are additional influence variables, e.g., the mean life span of a tumor cell T^* and the amount of tumor cell loss.

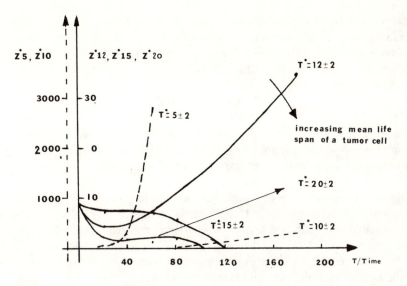

FIGURE 19. Time behavior of tumor cells $Z^* = f(T)$ with constant initial configuration (3 × 3 tumor nucleus in the center of the cell system 1; T1 = 10 ± 2 time units) and a variable mean life span T^* of the tumor cells. Tumor cell loss: 5%/time unit.

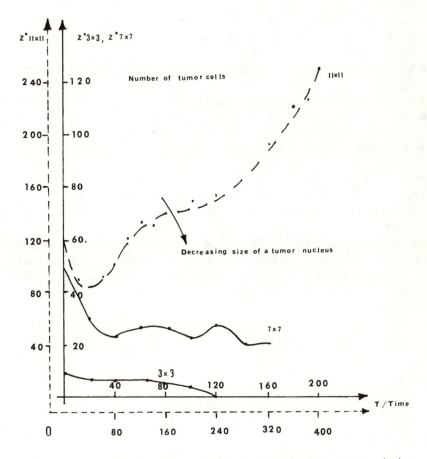

FIGURE 20. Time behavior of tumor cells $Z^* = f(T)$ with a variable tumor core size in the center of cell system 1 (mean life span of a cell: T1 = 10 ± 2; $T^* = 15 ± 2$ time units). Tumor cell loss: 5%/time unit.

Thus, this computer model enables the simulation of simple basic cases of the spread of cancer cells in tissue which in reality could hardly or not at all be tested. In the future it seems to be possible to partly substitute long and expensive biological experimental test series by simulation with the help of these models.

VII. CONCLUSION

Resuming the results of this chapter one can say that starting with studies on tumor diseases in view of unstable control loops, several models have been developed to describe the time behavior and the spatial structure of disturbed cell growth.

1. Stability conditions have been derived from a single-loop tumor growth control circuit.
2. A macro-model concerning the blood-forming process has been developed by which it is possible to describe the dynamics of different malignant blood diseases.
3. The macro-model of the blood forming process has been extended to a generalized compartment model. The time courses gained by simulations prove a great similarity to the kinetics of malignant cell renewal systems.
4. A grid model describing malignant cell growth in a tissue has been developed. The dynamics as well as the local position of the cells in an array have been studied by means of a minicomputer. The results are similar to morphological cuts gained by experiments.
5. In modeling the spread of cancer cells in tissue, special emphasis was given to:
    *   The existence of several cell systems with different mean life spans growing in competition
    *   The variability of the mean life spans of a cell and of the initial configuration of cell patterns
    *   The description of the cell-to-cell interactions
    *   The perturbations of normal cell growth by tumor cells and their elimination in medicine comparable with an irradiation or a removal by surgery
    *   The consideration of the loss of tumor cells

The results of these studies have helped to get a deeper insight into the structure and the time behavior of cell renewal systems. They are to stimulate further deliberated tests with cell cultures to give the hypotheses and simulation results of this contribution a full corroboration.

In the long run, the next task will be to extend the present study to a three-dimensional cell space using modified specifications. The following biological facts will be taken into consideration: the different ray sensibility in the single cell-cycle phases, the DNA repair, the immune response, and the multi-stage hypothesis of carcinogenesis from a normal cell into a cancer cell.

REFERENCES

1. **Eisen, M.,** Mathematical models in cell biology and cancer chemotherapy, in *Lecture Notes in Biomathematics,* Springer-Verlag, Basel, 1979.
2. **Bronk, B. V., Dienes, G. J., and Paskin, A.,** The stochastic theory of cell proliferation, *Biophys. J.,* 8, 1353, 1968.
3. **Jagers, P.,** *Branching Process with Biological Applications,* John Wiley & Sons, New York, 1975.

4. **Düchting, W.**, Krebs, ein instabiler Regelkreis, Versuch einer Systemanalyse, Kybernetik, Bd. 5, Heft 2, 1968, S.70—77.

5. **Jacobsen, L. O. and Doyl, M.**, *Erythropoiesis*, Grune & Stratton, New York, 1962.

6. **Covelli, V., Briganti, G., and Silini, G.**, An analysis of bone marrow erythropoiesis in the mouse, *Cell Tissue Kinet.*, 5, 41, 1972.

7. **Kennedy, B. J.**, Cyclic leukocyte oscillations in chronic myelogenous leukemia during hydroxyurea therapy, *Blood*, 35(6), 751, 1970.

8. **Stanton, M. F., Ed.**, Chalones: Concepts and Current Researches, National Cancer Institute Monograph, U.S. Government Printing Office, Washington, D.C., 1973.

9. Programmiersystem-Unterlagen, ASIM der Firma AEG-Telefunken, W. Germany, 1973.

10. **Dörmer, P.**, *Kinetics of Erythropoietic Cell Proliferation in Normal and Anaemic Man. A New Approach Using Quantitative ^{14}C-Autoradiography*, Gustav-Fischer-Verlag, Stuttgart, 1973.

11. **Düchting, W.**, Computer simulation of abnormal erythropoiesis — an example of cell renewal regulating systems, *Biomed. Tech.*, Bd. 21, Heft 2, 1976, 34.

12. **Mylrea, K. C. and Abbrecht, P. H.**, Mathematical analysis and digital simulation of the control of erythropoiesis, *J. Theoret. Biol.*, 33, 279, 1971.

13. **Düchting, W.**, A cell kinetic study on the cancer problem based on the automatic control theory using digital simulation, *J. Cybernet.*, 6, 139, 1978.

14. **Baserga, R.**, *Multiplication and Division in Mammalian Cells*, Marcel Dekker, New York, 1976.

15. **Kim, M., Bahrami, K., and Woo, K. B.**, Mathematical description and analysis of cell cycle kinetics and the application to Ehrlich ascites tumor, *J. Theoret. Biol.*, 50, 437, 1975.

16. **Gardner, M.**, On cellular automata, self-reproduction, the Garden of Eden and the game life, *Sci. Am.*, 224, 112, 1971.

17. **Lindenmayer, A.**, Developmental algorithms for multicellular organisms: a survey of L-systems, *J. Theoret. Biol.*, 54, 3, 1975.

18. **Zeleny, M.**, Self-organisation of living systems, *Intl. J. Gen. Syst.*, 4, 13, 1977.

19. **Düchting, W.**, A model of disturbed self-reproducing cell systems, in *Biomathematics and Cell Kinetics*, Valleron, A.-J. and Macdonald, P. D. M., Eds., Elsevier/North-Holland, Amsterdam, 1978, 133.

20. **Düchting, W.**, Spread of cancer cells in tissue: modelling and simulation, *Intl. J. Bio-Med. Comput.*, 11, 175, 1980.

Chapter 5

MODELING AND SIMULATION FOR DRUG DESIGN

E. V. Krishna Murthy

TABLE OF CONTENTS

"The best model of a cat is another cat or, better, the cat itself"

N. Wiener

I. ABSTRACT

Advances in computers and programming technology are beginning to make feasible the systematic or rational design of therapeutic drugs. The computer-based strategy which involves combinatorics, linear algebra, numerical analysis and statistics, along with organic and biochemical reaction mechanisms is applied at several levels:

1.   Prediction, selection, and design of a suitable drug — by correlating the physico-chemical properties or fragments or topological indices of the molecules with their biological activity and isolating those compounds which show promise in treating a disease and modifying these compounds by adding additional substituents and making finer adjustments (or modulation) in their biological activity.
2.   Understanding of the mechanism and response in the human body and the design of optimal dosage regimen — by studying the pharmacokinetical properties; this is done by simulating a suitable biodynamical system model of the human response through available in vivo data.

This article outlines the recent developments in the fast developing area of quantitative drug design, based on these concepts. Also, relevant information is provided regarding the tools needed for the purpose — choice of data structures, algorithms, and languages.

II. INTRODUCTION

A drug is a chemical substance that exerts a reproducibly observable effect on a biological system. This effect may be called biological activity or biological response and is usually dose-dependent. The dose is the quantity or concentration of the administered compound; the response is the observed effect which can be as objective as the percent inhibition of an enzymatic reaction or a change in the heart rate or as subjective as a change in the mood or posture of a test animal.

An important facet of modern medicinal chemistry is the quantitative drug design, wherein the chemist has to design a compound with a prescribed biological profile. While the achievement of this aim still appears to be far in the future due to the lack of our understanding the biological and disease processes at the molecular level, recently, computer-based strategy has been applied with reasonable success for the design of drugs at several levels:

A. Choice of a Drug

Ehrlich (1913; see Ariens[1]) was one of the earliest pioneers to study structure-activity relationships and hypothesize that the chemical properties of a drug definitely are a determinant factor for the biological action and thus, a relationship between chemical properties (molecular structure) and action or activity must exist. This is the basic principle used in all the investigations pertaining to the choice of a drug. Originally, this choice was made merely on a comparison of the well-known two-dimensional structural formula. However, soon it was realized that for some groups of drugs, causing certain types of action, namely anesthetic action, no structure action relationship could be detected on the basis of the structural formula. This necessitated the development of drugs based on correlating the physicochemical properties such as partition coefficient (lipophilicity), charge distribution constant, and steric parameters. Using these properties one can build models which predict the activity of a particular chemical compound. Basic to this approach is the search for specific subgroups and substructures, in particular chemical compounds and their physicochemical properties which are then correlated with their biological activities. This would evidently need the use of computers to analyze the substructures in a given compound and correlate their activities. We will discuss these in more detail in Section III.

B. Understanding Pharmacokinetics and Design of Dosage Regimen

This is carried out by studies of the pharmacokinetical properties of drugs. Here a suitable model (either a linear compartmental model or other nonlinear model) is chosen and this model is evaluated on the basis of its response to the data obtained in vivo by solving differential equations, using either an analog or hybrid computer or digital simulation. The model is then suitably modified for the determination of optimal dosage of the drug. After the dosage is determined it is again verified by simulation and comparison with the in vivo data. This, as well as a nonlinear model for optimization under use of multiple drugs are discussed in Section IV.

In Section V, we briefly comment on the choice of tools, namely data structures, algorithms, languages, and computers for this purpose.

III. MODELS FOR PREDICTION, SELECTION, AND MODULATION OF A DRUG

In order to establish rationale for the prediction and design of new candidate compounds with improved performance, it is necessary to derive a relationship between molecular structure or its suitable surrogate and the biological activity. Molecular structural description involves the number and nature of the atoms, bondings, and spatial relationship. For studying the structure-activity relationship, it is therefore necessary to quantitatively represent the molecular structure. The most complete description is obtained by using quantum mechanics. In particular, molecular orbital methods[2-4] have been used with success in structure-activity correlation; these, however, require enormous computation, since the information processed contains great detail such as the electron distribution probabilities and energies. Therefore, several alternative approaches which are simpler, have been proposed recently. These provide quantitative structure descriptors using the following three models:

1. Equation models based on physicochemical properties — Hansch-Free-Wilson[5] and Kubinyi[6] models
2. Topological fragment analysis and correlation model — Cramer's model;[7] this is based on relating the combination of contributions of the biological activities of structural components or substructures to the biological activity of a molecule
3. Topological connectivity indexes method — this is based on computing a topological index of a molecule which condenses the connectivity properties in the chemical graph associated with the structural formula — Hosaya, Wilcox, Randic, Kier models[8-12]

A. Equation Models Based on Physicochemical Properties

These are the most extensively used models due to their simplicity. The various physicochemical parameters like lypophilicity (partition coefficient), electronic properties, and steric factors are used here.[13] The most important among these is the lypophilicity. In relating biological activity mathematically to these properties of the molecule, one tries to construct a linear or nonlinear system of equations of the form

$$\text{Activity} = a_0 + \sum_{i=1}^{m} a_i x_i + \sum_{i=1}^{k} b_i x_i^2 \tag{1}$$

where x_i are the various properties and the coefficients a_i and b_i are determined by regression analysis. Equation 1 therefore permits prediction of the biological activity when x_i is known.

Procedures for obtaining the $(m + k + 1)$ coefficients $a_0 ..., a_m, b_1, ..., b_k$ require an experimentally determined biological activity and a group of $(m+k)$ properties for each of the n compounds chosen, which form a homologous series (viz., structurally related). Because of the relatively low precision of the biological data and the uncertainty of the applicability of the correlation model to the structure activity relations, good practice in drug design requires that $n \gg m + k$. The number of compounds in excess of the minimum $(n+1)$ required for analysis is called the number of degrees of freedom. In this context, it is important to note that the experimental variability in biological measurement is much larger than that of most physical or chemical measurements; hence, to obtain meaningful correlations it becomes necessary to obtain very reliable biological data with a well-defined confidence limit.

The coefficients a_i and b_k are usually obtained by using multiple regression or using generalized inverse methods.[14-15]

Usually the linear model turns out to be an oversimplification to handle more complex drug design problems. If the relationship is not linear and several variables are involved, it is necessary to use multivariable statistical analysis, e.g. Cluster analysis, a mathematical technique in which similarity is defined in terms of a distance between the points representing the objects (i.e., compounds) in multidimensional variable space.[16-17] We now describe the models that are widely used.

1. Hansch-Free-Wilson Parabolic Model

In this model, C is the concentration of the molecule necessary to give a defined response in terms of the partition coefficient P, σ, the Hammett constant of substituents, and E_s, the steric Taft parameters, thus:

$$\log \frac{1}{C} = a(\log P)^2 + b \log P + d\sigma + eE_s + f \tag{2}$$

The coefficients a, b, d, e, and f are computed by standard regression analysis.

Thus, here we assume that the biological activity is an additive property of the substituents that vary within the series. Analyses based on this model have been able to account for a substantial part of the variation of the biological activity in numerous compounds.

Rekker[18] explains the method for computing log P using the formula

$$\log P = \sum_{i=1}^{n} a_i f_i + \alpha \tag{3}$$

where f_i represent the hydrophobic fragmental constant; a_i is a numerical factor which varies with the incidence of a given fragment in the structure, and α is a correction factor due to anamolies present in the structure. This formula helps in computing log P of a molecular structure from the individual f_i values of the fragments and their spatial configuration.

2. Kubinyi Model[6]

Kubinyi observes that the above parabolic model is not precise as another nonlinear model which he constructs based on the investigation of the drug transport through a three-compartment model for the drug-receptor system. This model is given by:

$$\log \frac{1}{C} = a \log P - b \log (\beta P + 1) + d\sigma + eE_s + f \tag{4}$$

Here the coefficients a, b, d, e, and f are determined by linear multiple regression, while β is a nonlinear term determined by stepwise iteration or Taylor series iteration.

Kubinyi's model has the following interesting properties:

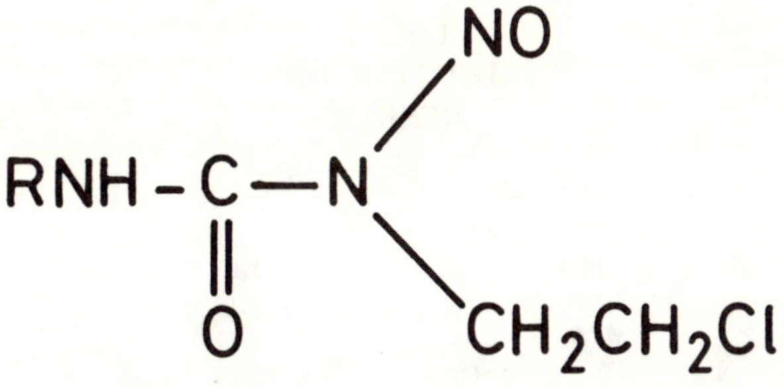

FIGURE 1. Nitroureas series.

1. The residuals resulting from this model are randomly distributed, unlike the parabolic model where the residuals show a regular pattern.
2. The models fit the observations very closely to the true data.

3. Examples

As an example of correlation results involving physicochemical properties, we consider a series of nitroureas (Figure 1) which delay the growth of a solid tumor, the Lewis lung carcinoma, in mice.[13] Here $\log 1/C = -0.08(\log P)^2 + 0.14 \log P + 1.23$.

For a large variation among the substituent groups, R, this equation holds.

The following example illustrates the predictive drug design based on the above model.[5] Table 1 shows the log P values of some of the most potent general anaesthetics which at one time or other have been clinically important. These results suggest a close relationship between the mechanism of action of barbiturates and gaseous anaesthetics.

B. Topological Fragment Analysis and Correlation Model

A severe limitation of the Hansch-Free-Wilson model is that it is restricted to structurally closely related compounds and it is inappropriate for (1) correlation of data where compounds fall into different structural series or into no series at all and (2) prediction of active compounds outside a structural class of established biological interest.

Therefore, Cramer et al.[7] developed a substructural analysis method; here, the biological activity of a molecule is related to the combination of contributions of the biological activities of structural components or substructures and their intra- and intermolecular interactions. Since a very large body of information will be generated (by the resulting combinatorial operations), even with the most modest-sized molecule, we need to make some simplifying assumptions such as:

1. The probability of a given biological activity can be usefully approximated by a first order analysis of substructural contributions (i.e., ignoring interactions).
2. The contribution of a given substructure to the probability of activity can be obtained from data on previously tested compounds containing that substructure.

With these two assumptions and a topological structure or substructure coding schemed, the basic functional groups, rings, chains, inorganic moieties, and other frequent combinations are recognized. Such a substructure coding scheme is naturally based on recognizing the subgraphs in a graph. Algorithms for automatic recognition of these are currently being developed by chemical-information scientists. Having recognized these basis fragments, the next step is to prepare a substructure "experience table" summarizing the data. A substructure activity frequency (SAF) defined for each substructure as (A/T) and the ratio of the number of active compounds (A) containing that substructure to the number of tested compounds

Table 1
PREDICTIVE DRUG
DESIGN

Anesthetic	Log P
Chloroform	1.97
Ethrane	2.10
Methoxyflurane	2.21
Trichloraethylene	2.29
Halothane	2.30

(T) containing the substructure represents the contribution that substructure can make to the probability of a compound being active. The experience table contains n(Say) SAFs corresponding to complete set of n substructures (fragments) previously recognized and coded in the test compounds.

Then we compute for each compound a "mean substructure activity frequency" (MSAF), the arithmetic mean of the SAF values of the substructure present in that compound. Example:

Consider the structure given in Figure 2, here MSAF is computed based on substructural analysis among the 770 compounds tested for anti-arthritic/immunoregulatory activity (see Table 2).

The 770 compounds were then marked in descending MSAF value. It was found that those with higher MSAF values were active more frequently than those with low MSAF values.

This technique has been successfully used to predict the probability of certain types of activity.

C. Topological Connectivity Index Method

We already mentioned that the structure-activity relationship needs a complete quantitative description of the molecular structure. In seeking simple alternatives to the use of quantum chemical methods, other approaches for quantification have been attempted. The most common of these, namely, the Hansch-Free-Wilson model uses the physicochemical properties. These can be called property-activity relationship schemes. As these methods develop an indirect relationship between activity and structure, they do not permit an unambiguous relationship between salient structure characteristics and the activity which they influence. Therefore, several attempts have been made to provide quantitative structure descriptors based on indexes derived from topology and geometry. Hence no physicochemical mechanism is presupposed in this approach. The basis for this approach is the fact that physical properties are closely correlated with the topological features of a molecule.

Recent development in the representation of chemical species have shown that a topological index which condenses the connectivity property associated with a chemical graph (structural formula treated as an abstract graph, identifying atoms as nodes and bonds as edges; we refer to the book by Harary,[19] for the concepts of graph theory) plays a great role in structure activity relationship. Several topological indexes can be defined. A potential use of these indexes is that one might find a correlation between the biological activity and these indexes as with the case of the physical and chemical properties. We now describe some of these indexes and their applications.

1. Hosaya Index[8-10]

The Hosaya index, H, is defined by

$$H = \sum_{k=0}^{m} p(G, k) \tag{5}$$

83

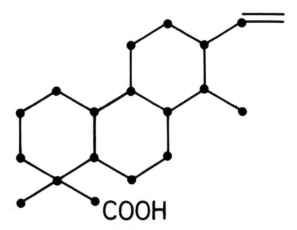

FIGURE 2. Sub-structural analysis method.

where p(G, k) is the number of ways in which k edges of the chemical graph can be chosen so that no two of them are directly bonded and the summation extends over all the m edges of the graph.

It has been found that the boiling point of organic substances is more or less proportional to H (see Table 3).

This promising correlation obtained by using the Hosaya index stimulated further interest among the scientists which led to te suggestion for other indexes given below.

2. Wilcox Index[9]

Wilcox suggested a simple connectivity index as a measure of molecular branching while calculating molecular π − orbital energies. This index, W, called the Wilcox index is defined by

$$W = \sum_{i=1}^{n} d_i^2 \qquad (6)$$

where d_i = degree of each node or the number of attachments to each atom, n = number of atoms. This is not used in practice.

3. Randic Index[9-12]

The Randic index, R, is defined by

$$R = \sum_{i,j} (d_i d_j)^{-1/2} \qquad (7)$$

where d_i and d_j are the degrees of all pairs of atoms i and j connected by a bond and the summation is taken over all such pairs.

A modified higher order form of the Randic index is used by Kier.

4. Kier Index[11-12]

Kier has used higher order molecular connectivity indexes for correlating the physical and biological properties. The basic Randic index, R, is based upon a single bond; higher indexes are calculated by summing indexes based upon two, three, or more adjacent edges in the chemical graph.

Thus, the pth order Kier's index K_p is given by:

Table 2
COMPUTATION OF MSAF BY SUBSTRUCTURAL ANALYSIS

Fragment	Tested(T)	Active(A)	SAF(A/T)
Aliphatic COOH	108	23	0.212
Phenanthrene ring system, 1 unsaturation	1	1	1.000
Carbon chain as functional group	198	47	0.247
Methyl chain	309	83	0.268
Quarternary carbon	78	18	0.230
Three rings	174	45	0.258
			2.215

Note: MSAF = 2.215/6 = 0.37

Table 3
HOSAYA INDEX AND BOILING POINT

Organic Compound	H	Boiling point in °C
Methane	1	− 162.0
Ethane	2	− 88.6
Propane	3	− 42.2
Isobutane	4	− 11.7
n-butane	5	− 0.5
n-pentane	8	36.1
n-hexane	13	68.7

$$K_2 = (d_i d_j d_k)^{-1/2} \qquad (8)$$

$$K_3 = (d_i d_j d_k d_m)^{-1/2} \qquad (9)$$

$$K_p = (d_i d_j \ldots)^{-1/2}$$
$$(p + 1) \text{ terms} \qquad (10)$$

The second order index K_2 represents a summing of three atom fragments where the summation is taken over every possible pair of adjacent bonds. Similarly K_3 represents a summing for each and every three-bond path through the molecule, each term being the reciprocal of the square root of a product of four degree values.

Kier also suggested another index V_p called p^{th} order Valency index defined by

$$V_p = \sum_i (\pi_j \Delta_j)^{-1/2}, j \neq i \qquad (11)$$

here the Valency Δ_j replaces the degree d_i on index K_p.
Example:

1.   Consider the graphs in Figure 3 (a and b).
     In Figure 3(a)

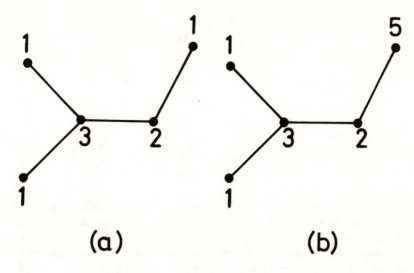

FIGURE 3. Calculation of Kier indexes.

$$W = d_i^2 = (1+1+9+4+1) = 16$$

$$K_0 = d_i^{-1/2} = (1+1+\frac{1}{\sqrt{3}} + \frac{1}{\sqrt{2}} + 1) = 4.285$$

$$K_1 = \left(\frac{1}{\sqrt{3}} + \frac{1}{\sqrt{3}} + \frac{1}{\sqrt{6}} + \frac{1}{\sqrt{2}}\right) = 2.270$$

$$K_2 = \left(\frac{1}{\sqrt{3}} + \frac{1}{\sqrt{6}} + \frac{1}{\sqrt{6}} + \frac{1}{\sqrt{6}}\right) = 1.802$$

In Figure 3(b)

$$V_0 = \left(1 + 1 + \frac{1}{\sqrt{3}} + \frac{1}{\sqrt{2}} + \frac{1}{\sqrt{5}}\right) = 3.732$$

$$V_1 = \left(\frac{1}{\sqrt{3}} + \frac{1}{\sqrt{3}} + \frac{1}{\sqrt{6}} + \frac{1}{\sqrt{10}}\right) = 1.879$$

$$V_2 = \left(\frac{1}{\sqrt{3}} + \frac{1}{\sqrt{6}} + \frac{1}{\sqrt{6}} + \frac{1}{\sqrt{30}}\right) = 1.576$$

2. We also now give an example to indicate how the vapor toxicity of alcohol (against red-spider)[12] varies with K_1 (Table 4).

IV. MODELS FOR DOSAGE REGIMEN

A. Pharmacokinetic Model

The selection of the most suitable drug as well as the rational therapy design are greatly dependent on the pharmacokinetical properties of the dosage form. The passage of a drug from the point of application to the site of action is very complicated. Many processes such as dissolution, absorption, distribution, metabolism, and finally elimination are involved. Due to the complexity of each of these processes, we can only model these using some simple concepts: (1) a mathematical formulation which is a near true representation of the natural phenomena as observed from the in vivo data and (2) correct numerical solution of the dynamics of the model.

Table 4
KIER INDEX AND TOXICITY

Compound	K_1	Observed toxicity	Calculated toxicity
Methanol	1.0	2.8	2.75
Ethanol	1.41	3.0	3.03
Propanol	1.91	3.3	3.37
Isopropanol	1.75	3.26	3.26
Isobutanol	2.270	3.72	3.60
Pentanol	2.9	4.09	4.04

For our purpose the most suitable model is the compartmental model,[20] which uses the analogy between the volume of a fluid in a leaky tank and the concentration of the drug in particular body compartments. In the compartmental model for studying a biodynamic system, we decompose the system into certain elementary parts called compartments. Each compartment denotes a particular body part, in which we assume that the concentration of the relevant drug is uniformly distributed. These compartments are then assumed to be interconnected in a network form with a single input and an output stage with some feedback between certain compartments. The complete model of the biodynamical system is then the aggregate network. If the drug is transferred from one compartment to another, the concentration in the second will rise to a maximum value and then decay (which corresponds to connecting two leaky tanks). This model provides an analogy for absorption, metabolism, and loss of the drug from the body although it may not be a true representation. The simplicity of this model permits us to choose:

1.   A linear compartmental model in which the system is described by a set of first order ordinary differential equations
2.   A nonlinear model in which the system is described by a set of nonlinear differential equations

We will only be concerned here with the former model due to its simplicity.

In Figure 4, we represent a typical biodynamic model; here k_1, k_2, k_3, and k_4 are rate constants which represent the rate of change of concentration from one compartment to another.

The relevant system of differential equations are

$$\frac{dA}{dt} = - k_1 A$$

$$\frac{dB}{dt} = k_1 A - k_2 B - k_3 B + k_4 D$$

$$\frac{dC}{dt} = k_2 B$$

$$\frac{dD}{dt} = k_3 B - k_4 D \qquad\qquad (12)$$

Our aim is to determine the various concentrations A, B, C, and D after a time lapse, T, given that at zero time the initial values are A_0, B_0, C_0, D_0 and k_1, k_2, k_3, and k_4 are known.

In the physiological sense each member of this compartment stands for either a tissue with equal distribution properties, or an organ or a body fluid like plasma or urine. In the pharmacokinetical sense, each compartment is defined by a volume and the input/output rates.[21]

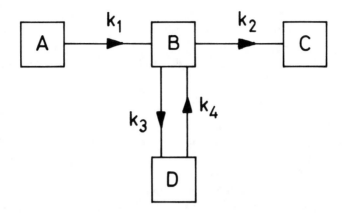

FIGURE 4. Compartmental physicokinetic model.

For medicines given orally the preceding model in Figure 4 may have the following interpretation. A is the gastrointestinal tract, B is the plasma, C is urine, D is a tissue. k_1, k_3, k_4 are relevant absorption rate constants and k_2 the elimination rate constant.

It is obvious that the solution to the system of differential equations (12) at various time intervals would tell us the drug concentration in various compartments at these intervals. If we know the minimum level of concentration of the drug for curing a given disease and the unfavorable high dosage, we can design a suitable dosage regimen (which prescribes multiple dosing at various intervals) for the drug. In other words, the solution of the differential equations tells us how to select the dosage quantity and interval to keep the drug concentration in the plasma between certain specified therapeutic limits.

It is clear that the solution to Equation 12 is an exponential characterized by an initial value and rate constant parameter indicating the rate of decay.

A straightforward method of simulating this system is to start at time zero and increment time in small steps of Δt, assuming that the variables A, B, C, and D remain unaltered during this time interval, and change instantaneously at the end of the step.

Thus, the quantity of A, B, C, D at the end of one such step is given by:

$$A(t + \Delta t) = A(t) + \frac{dA}{dt} \Delta t \tag{13}$$

If we want to simulate the system over a period, T, then we divide T into N equal intervals such that

$$N \Delta t = T \tag{14}$$

At time zero, we know, A(0), B(0), C(0), and D(0) and from these initial values and values of k_1, k_2, k_3, and k_4 we can compute the new values $A(\Delta t)$, $B(\Delta t)$, $C(\Delta t)$, and $D(\Delta t)$. For example

$$
\begin{aligned}
A(\Delta t) &= A(0) - k_1 A(0) \cdot \Delta t \\
B(\Delta t) &= B(0) + [k_1 A(0) - k_2 B(0) - k_3 B(0) \\
&\qquad + k_4 D(0)] \Delta t \\
C(\Delta t) &= C(0) + k_2 B(0) \Delta t \\
D(\Delta t) &= D(0) + [k_3 B(0) - k_4 D(0)] \Delta t
\end{aligned} \tag{15}
$$

Using these we can compute A, B, C, and D at the next time interval and so on for the N steps at the end of which we will have the desired trajectory for each one of the variables.

The above digital computer simulation procedure uses Euler's interpolation formula.[22] It is possible to use a more sophisticated method like the Runge-Kutta method for this purpose.[22] Also, an analog (hybrid) computer can be used.

In the above example, we described a model with a few compartments. In some cases, it may be necessary to use a multicompartmental system with complicated feedback or reversible paths. These may lead to large nonlinear differential systems. As the system of equations become larger and larger and the time constants differ widely the system of differential equations could turn out to be very stiff, demanding an excessive amount of computation time to achieve a desired accuracy. In such cases, one has to use special software; see Shampine and Gear[23] and Shampine and Watts.[24]

The above example is merely illustrative of the approach. In a practical situation the extent of errors involved should be determined by simulation by adding random errors in data.

Also it may happen that in many cases a multicompartmental model is required. This model may be modified in various ways by lumping compartments, allowing or disallowing exchange among specific compartments and so on.

Karba et al.[25] recently discussed a compartmental model for the dosage regimen for a bronchial dilator. Their results are quite promising.

B. Nonlinear Optimization for Prescription

For certain specific illness and conditions, several compounds or drugs that show promise have to be administered as a suitable combination. The dosage as well as the relative ratios of the compounds being subject to therapeutic limits have to be chosen to minimize the number of days of treatment or administration of the drug. This turns out to be a nonlinear constrained optimization problem.

Suppose n compounds C_1, C_2, ... C_n are chosen for this purpose and there is a therapeutic range x_i (in milligrams) for each C_i ($i = 1$ to n) given by

$$\alpha_i \leq x_i \leq \beta_i \tag{16}$$

and we have other constraints such as certain linear combinations of the drugs C_i should remain within a certain range as given by

$$\gamma \leq Ax \leq \beta \tag{17}$$

where A is an (m × n) numerical matrix prescribing linear combinations of the x_i (component of x) that should be between γ_i (component of γ) and β_i (component of β). Then we need to construct a nonlinear model for minimizing, y, the number of days of treatment or administration of the drugs, given by

$$y = a_0 + a_1 x_1 + a_{11} x_1^2 + a_{12} x_1 x_2 \ldots + a_{ij} x_i x_j \tag{18}$$

where a_{ij}'s are known coefficients empirically determined.

This can be done either by nonlinear optimization techniques or by multistage Monte Carlo method; see Conley.[26]

V. TOOLS FOR MODELING AND SIMULATION

In the last two sections, we described several models for use in drug design. These models and simulations essentially involve computational algorithms falling under the following three categories:

1. Numerical and statistical algorithms: polynomial fitting, optimization, regression, and correlation

2. Combinatorial and graph algorithms: substructure analysis, topological index computation, searching, and ordering.

3. Continuous simulation algorithms: solution of biodynamic compartmental models involving differential equations

Each of these three categories require the use of different kinds of data structures and languages for writing suitable programs.

In addition, all of these would need computer manipulation of chemical structural records as well as records pertaining to the physicochemical-biological properties of the compounds. Therefore, a proper data base organization is needed.

Current day developments in programming languages make it possible for us to implement or simulate all the above models in a digital computer with great ease. For instance, PASCAL or LISP can be used for implementing graph algorithms needed for fragmenting and substructure analysis of the chemical graphs; see Wirth,[27] Even,[28] Horowitz and Sahni,[29] and Grogono.[30]

The dynamic models can be simulated either with an analog, hybrid, or digital computer. When a digital computer is to be used, proper choice of a simulation language must be made. Continuous-system simulation languages are designed to facilitate modeling and solution of dynamic problems formulated in terms of ordinary and partial differential equations. Each language has a different modeling orientation, range of capabilities, and applications.[31]

The languages such as MIDAS, MAD BLOCK, and CSMP 1130 are block-oriented and so have remarkable simplicity and naturalness to simulate a dynamical system. If an interactive set-up is available, a small digital computer can effectively serve as an analog computer. However, block-oriented languages are not flexible because of the small vocabulary. Therefore, expression-oriented languages have been developed which handle directly the differential equations representing the model.

MIMIC is a language in which FORTRAN-like statements can be used. The solution of differential equations is obtained using a fourth order Runge-Kutta method.

DAREP (Differential Analyzer Replacement Portable) is another language which is nearly FORTRAN except for a small set of machine-dependent routines.

CSMP/360 and CSMP-III can handle either expressions or block-oriented descriptions. They are all preprocessors with FORTRAN as an intermediate language.

CSSL-III is a language recommended by a simulation software committee; this language has a powerful capability. It has error control and debugging facilities. Also, several choices of integration methods can be made.

Finally, it is necessary to touch upon the role of microcomputers as a simulation tool. These turn out to be very convenient for simulating the compartmental kinetics model.[32] Many microcomputers currently have graphic facilities available which can plot the variables as a function of time. These will serve as a valuable tool in the clinical laboratory.

VI. CONCLUDING REMARKS

The essential purpose of our survey here is to indicate the role of computers, combinatorics, and statistics as emerging essential tools for the design of drugs. The methods described in this survey are the current preliminary attempts to construct quantitative measures for the biological activity of a molecule based on the assumption that the function of "a whole" can be described in terms of the contribution from its "parts". While these techniques by themselves do not lead to a fool-proof design of drugs, they are very useful as preliminary steps for the selection of a suitable group of drugs for testing.

Also, the simulation of the pharmacokinetical model is very useful for determining the dosage and response of individual patients.

In summary, computers have a great role to play in quantitative drug design and dosage regimen.

REFERENCES

1. **Ariens, E. J.,** Excursions in the field of SAR. A consideration of the past, the present and the future, in *Biological Activity and Chemical Structure,* Buisman, K. Ed., Elsevier, New York, 1977, 1.
2. **Kier, L. B.,** *Molecular Orbital Theory in Drug Research,* Academic Press, New York, 1971.
3. **Kier, L. B.,** *Molecular Orbital Studies in Chemical Pharmacology,* Springer-Verlag, New York, 1971.
4. **Purcell, W. P., Bass, G. E., and Clayton, J. M.,** *Strategy of Drug Design,* John Wiley & Sons, New York, 1973.
5. **Hansch, C.,** On the predictive value of QSAR, in *Biological Activity and Chemical Structure,* Buisman, K., Ed., Elsevier, New York, 1977, 47.
6. **Kubinyi, H.,** Non-linear dependence of biological activity on hydrophobic character: the bilinear model, in *Biological Activity and Chemical Structure,* Buisman, K., Ed., Elsevier, New York, 1977, 239.
7. **Cramer, R. D., Redl, G., and Berkoff, C. E.,** Substructural analysis: a novel approach to the problem of drug design, Rep. of Technol. Assessment Divis., Smith, Kline and French Laboratory, Philadelphia, 1976.
8. **Hosaya, H.,** Topological index as a sorting device for coding chemical structures, *J. Chem. Document.,* 12, 181, 1972.
9. **Evans, L. A., Lynch, M. F., and Willett, P.,** Structural search codes for on-line registration, *J. Chem. Inf. Comp. Sci.,* 18, 146, 1978.
10. **Rouvray, D. H.,** The search for useful topological indices in chemistry, *Am. Sci.,* 61, 729, 1973.
11. **Kier, L. B. and Hall, L. H.,** *Molecular Connectivity in Chemistry and Drug Research,* Academic Press, New York, 1977.
12. **Kier, L. B. and Hall, L. G.,** The nature of structure-activity relationship and their relation to molecular connectivity, *Eur. J. Med. Chem.,* 12, 307, 1977.
13. **Redl, G., Cramer, R. D., and Berkoff, C. E.,** Quantitative drug design, *Chem. Soc. Rev.,* 3, 273, 1974.
14. **Albert, A.,** *Regression and the Moore-Penrose Inverse,* Academic Press, New York, 1972.
15. **Rao, C. R. and Mitra, S. K.,** *Generalized Inverse of Matrices and its Applications,* John Wiley & Sons, New York, 1971.
16. **Sneath, P. H. A. and Sokal, R. R.,** *Numerical Taxonomy,* W. H. Freeman, San Francisco, 1973.
17. **Hawkes, J. G.,** *Chemotaxonomy and Serotaxonomy,* Academic Press, New York, 1968.
18. **Rekker, R. F.,** The hydrophobic fragmental constant, in *Biological Activity and Chemical Structure,* Buisman, K., Ed., Elsevier, New York, 1977, 231.
19. **Harary, F.,** *Graph Theory,* Addison-Wesley, Reading, Mass., 1969.
20. **Riggs, D. S.,** *The Mathematical Approach to Physiological Problems,* M.I.T. Press, Cambridge, Mass., 1963.
21. **Hemker, H. and Hess, B.,** *Analysis and Simulation of Biochemical Systems,* Elsevier, New York, 1972.
22. **Young, D. M. and Gregory, R. T.,** *A Survey of Numerical Mathematics,* Addison-Wesley, Reading, Mass., 1973.
23. **Shampine, L. and Gear, C. W.,** A user's view of solving stiff ODE, *SIAM Rev.,* 21, 1, 1979.
24. **Shampine, L. and Watts, H. S.,** Software for ODE, in *Sources and Development of Mathematical Software,* Cowell, W. R., Ed., Prentice Hall, Englewood Cliffs, N.J., 1981.
25. **Karba, R., Bremsak, F., Kozjek, F., Mrhar, A., and Matko, D.,** Hybrid simulation in drug pharmacokinetics, in *Simulation of Systems '79, Proc. of 9th IMACS Congress,* Dekker, L., Savastano, G., and Vansteenkiste, G. C., Eds., North Holland, Amsterdam, 1976.
26. **Conley, W.,** Making a prescription drug, *Int. J. Math. Educ. Sci. Technol.,* 12, 425, 1981.
27. **Wirth, N.,** *Algorithms + Data Structures = Programs,* Prentice Hall, Englewood Cliffs, N.J., 1976.
28. **Even, S.,** *Graph Algorithms,* Computer Science Press, Potomac, Md., 1979.
29. **Horowitz, E. and Sahni, S.,** *Fundamentals of Computer Algorithms,* Computer Science Press, Potomac, Md., 1978.
30. **Grogono, P.,** *Programming in PASCAL,* Addison-Wesley, Reading, Mass., 1980.
31. **Korn, G. A. and Wait, J. V.,** *Digital Continuous Systems Simulation,* Prentice Hall, Englewood Cliffs, N.J., 1978.
32. **Randall, J. E.,** *Microcomputers and Physiological Simulation,* Addison Wesley, Reading, Mass., 1980.

Chapter 6

COMPUTER MODELING IN CARDIOVASCULAR RESEARCH: GUYTON MODELS AND ESSENTIAL HYPERTENSION

Nguyen Phong Chau and Michel E. Safar

TABLE OF CONTENTS

I. BASIC KNOWLEDGE ABOUT HYPERTENSION

A. Definition

Hypertension is, by definition, a pathology in which arterial blood pressure is higher than normal. However, in a wide population, the distribution of arterial blood pressure is of Gaussian type. Thus, the choice of a precise pressure level to define hypertension is arbitrary. The most widely accepted criteria are

- Normal blood pressure = systolic blood pressure below 140 mmHg and diastolic blood pressure below 90 mmHg
- Sustained hypertension = systolic blood pressure above 160 mmHg and diastolic blood pressure above 95 mmHg
- Borderline hypertension = blood pressure not included in the two preceding cases

All epidemiological studies showed that blood pressure is highly correlated with cardiovascular morbidity and mortality. However, the importance of this finding was recognized only when, in the last decades, the effectiveness of antihypertensive therapy has been demonstrated; antihypertensive drugs reduce significantly cardiovascular morbidity and mortality. Since high blood pressure affects about 10% of the adults, such results point to the social importance of the hypertensive disease.

B. A Rapid Overview on Hypertension

Our knowledge about hypertension is based both on clinical investigation and animal experiments.

Animal studies were devoted to the discovery of "causes" of hypertension and mainly to the role of the kidney. A classical example was the observation of Goldblatt that clamping a renal artery raised blood pressure. Another example was the finding of Page and Braun-Menendez that a polypeptide, now called angiotensin, had extremely powerful vasoconstrictor effects on arterioles and veins. Angiotensin results from the action of a renal enzyme, renin, on its hepatic substrate, angiotensinogen. Angiotensin constricts the vessels, increases their resistance to flow and consequently raises blood pressure. Other examples of renal or adrenal causes of hypertension have been discovered and are listed in Table 1. Some of them can be cured by surgery. When a cause is discovered, hypertension is called "secondary" (to the cause which has been found).

However, in a great majority (more than 90%) of cases of high blood pressure *observed in man*, no cause of hypertension could be identified. Hypertension is then called "essential", i.e., of unknown cause.

Essential hypertension might be a hereditary disorder, and animal models of genetic hypertension, such as spontaneously hypertensive rats, were developed. On the other hand, precise and sophisticated measurements of blood pressure, cardiac output, blood volume, extracellular fluid volume, total body water, renin, angiotensin, aldosterone, etc., raised the question of the role of all these parameters in the genesis and development of high blood pressure.

It appeared that hypertension is essentially a long-term and multiparametric disease. The occurrence of high blood pressure provokes several modifications in the global behavior of the circulatory system. Many physiological functions are changed, either as impairments due to high blood pressure, or as adaptation of the organism to live with the hypertensive state.

C. Experimental Hypertension and Guyton's Models

From animal studies on the genesis and development of hypertension, four dominant groups of factors emerged: (1) hemodynamics (blood pressure, blood flow, and resistance to flow), (2) fluid volumes and the kidney, (3) neuro-hormonal parameters such as renin, angiotensin, etc., and (4) the sympathetic nervous system.

Table 1
NORMAL LEVELS OF THE VARIABLES AND NORMAL VALUES OF THE COEFFICIENTS

UO	= 1 mℓ/min		ECFV	=	15 ℓ
VAS	= 1 normalized unit		BV	=	5.25 ℓ
TPR	= 0.01323 mmHg·min/mℓ		AR	=	0.010569 mmHg·min/mℓ
MSP	= 7 mmHg		AP	=	90 mmHg
AUM	= 1 normalized unit		CO	=	6.8 ℓ
I	= 1 mℓ/min				
a_1	= 0.06		b_1	=	-4.4
a_2	= 0.35		b_2	=	0
a_3	= 9.52		b_3	=	-42.98
a_4	= -0.59		b_4	=	4.4152
a_5	= 0.45		b_5	=	-3.06
a_6	= -0.002		b_6	=	1.18
a_7	= -0.005		b_7	=	1.45
k_1	= 0.010569		k_2	=	0.4052
k_3	= 0.2304		k_4	=	1
α	= 0.75		β	=	8/31
γ	= 1/31		VRES	=	0.002666

Note: UO = urinary output, VAS = vasculature index, TPR = total peripheral resistance, MSP = mean systemic pressure, AUM = autonomic multiplier, ECFV = extracellular fluid volume, BV = blood volume, AR = arterial resistance, AP = arterial pressure, CO = cardiac output, VRES = venous resistance.

In the last 20 years, considerable efforts have been made by Guyton and collaborators to construct, develop, and test global models of blood circulation and hypertension. Some of them have now reached a high degree of complexity.[2,3]

D. Main Features of Guyton's Models

The "big" Guyton models include almost all the variables physiologically related to blood circulation. About 400 parameters were analyzed. They were interrelated by definition equations, or by classical equations borrowed from our knowledge in physics, mechanics, or chemistry. However, a number of them, called "physiological function blocks", have been directly assessed by experimentation on the living animal.

From the mathematical viewpoint, a Guyton model is an open and stable dynamic system. In the absence of perturbation, with a constant input, the system stays at a steady state. Under the effect of a perturbation, the system undergoes a transient state. Perturbations to the system can be classified into 3 types:

1. *A shock,* i.e., a sudden change of initial conditions, brings the system abruptly out of its steady state. Then, because of stability, the system will return to its initial steady state.
2. *An infusion,* i.e., a sustained change of input to the system, moves the system towards a new steady state compatible with the new input level. If infusion ceases, the system will return to its initial steady state
3. *An impairment,* i.e., an irreversible change of an internal function, also moves the system towards a new steady state compatible with impaired internal function.

In most studies of animal hypertension except those of spontaneously hypertensive rats, each of the above perturbations was provoked and the resulting hemodynamics were followed as functions of time. In that manner, the function blocks were quantified, the causes of hypertension were localized, and the behavior of the organism at the onset of hypertension was investigated.

E. Human Study of Hypertension

The above animal method cannot be applied to man. In man, only a limited number of physiological components are measurable. Only some of them can be artificially changed over a narrow range and the consequences of these changes followed with time. Perturbations capable of producing sustained hypertension cannot be realized.

However, in man, some measurements (blood pressure, blood flow, fluid volumes, different renal factors, etc.) can be performed in a large number of individuals, mostly patients for whom the measurements are required for clinical investigations. Human studies of hypertension have mainly been based upon these measures.

It is important to understand that clinical data are of different nature compared to those of animal experimentation. For example, the set of data (blood pressure, heart rate, cardiac output, blood volume, extracellular fluid volume, renal blood flow, etc.)[4-6] was obtained under the following conditions: the patients were hospitalized and stayed in equilibrated sodium balance for several days. They were either untreated or had therapy discontinued for several weeks. There was no reason to believe that any patient had an acute perturbation or damage just before hospitalization. The measurements were performed with the patients at rest. For all of these reasons, it is clear that only specific steady states of the patient's hemodynamics have been determined. As a consequence, any set of data obtained in comparable conditions as above consisted of different steady states of different patients with different degrees of pathology.

Clinical data were submitted to statistical analyses. When a group of normotensives was compared to a group of hypertensives, mean levels of (and correlations between) some variables were different.[4] These differences clearly indicated that hemodynamic regulations in the hypertensives were abnormal. However, for many reasons, the physiological interpretations of the results were not easy: Firstly, only steady states were studied in man. Deviations of steady states cannot be compared with short-term experimental relationships. Secondly, full measures, including determinations of blood pressures, fluid volumes, etc., cannot be repeated several times in the same subjects (generally the determinations were performed only once in a patient). Thirdly, when hypertension is detected, the pathology has probably occurred for months or years. The mechanism(s) initiating the pressure rise cannot be seen. Fourthly, during the persistance of hypertension, several physiological functions have been modified, either as impairments due to high blood pressure or as adaptation of the organism to live with the hypertensive state.

F. The Use of Models in Human Hypertension

Human hypertension can be investigated only in the framework of a global model. As such a model cannot be directly assessed in man, the only way to go ahead is to extrapolate animal models to man. The extrapolation raises of course several problems. Some of them will be discussed in this chapter.

To be specific, we chose a model of blood circulation and a set of data. The chosen model was that published by Guyton and Coleman.[7] The chosen data were those published by Safar et al.[4,5]

Section II. recalls the main features of Guyton-Coleman model. To prepare the use of the model in human hypertension, the steady states and deviations of steady states were carefully examined. Section III. recalls the data of Safar et al. To facilitate the matching of data with model predictions, the moving-mean procedure was used to delineate the deviations of the observed steady states following the pressure rise. Section IV. is an attempt to match the deviations of steady state in models. The matching procedure disclosed some basic facts on the progress of human hypertension.

II. THE 1967-GUYTON-COLEMAN MODEL

A. Basic Structure

The heart is formed by two pumps (Figure 1): blood coming from the systemic circulation enters the right atrial (A), is driven by auricular muscle contraction to the right ventricle (B) where the right pump propels it to the lung (C). Blood coming from the lung enters the left atrial (D), is driven to the left ventricle (E) where the left pump propels it to the systemic circulation. The whole systemic network consists essentially of a "high pressure" zone (aorta, large arteries, arterioles) with a small volume, a small capacitance but a high resistance and a "low pressure" zone (veinules and large veins) with a large volume, a great capacitance, but an almost null resistance.

The basal structure of the Guyton-Coleman model includes: (1) an arterial compartment with an unstressed volume VO_a, a capacitance C_a, an arterial resistance R_a and (2) a venous compartment with an unstressed volume VO_v, a capacitance C_v, and a venous resistance R_v. Filled with the circulating blood, the two compartments had stressed volume V_a and V_v. Blood pressures in the compartments were denoted by P_a (or AP) and P_v, respectively. P_{ra} designated right atrial pressure and cardiac output (CO), the blood flow through the system. All variables were mean values over a period.

Let $BV = V_a + V_v$, $BVO = VO_a + VO_v$, Guyton defined the "mean systemic pressure", MSP, by the equation:

$$MSP = (BV - BVO)/(C_a + C_v)$$

and derived his celebrated formula:

$$CO = (MSP - P_{ra})/R_{vr}$$

where R_{vr}, the "resistance to venous return" was defined by:

$$R_{vr} = R_v + C_a R_a/(C_a + C_v)$$

On the above basic structure, Guyton and Coleman grafted several physiological function blocks, most of them established from experimentation in dogs.

B. The Physiological Function Blocks

1. The Kidney

The salt and water excretory function of the kidney was quantified by the so-called "pressure-diuresis" curve. The kidney excretes more or less water and salt following the blood pressure level. The experimental curve is depicted in (Figure 2, curve 1).

2. The Partition of Fluid Volumes

The partition of fluids between the intravascular and extracellular compartments was described by an experimental curve (Figure 2, curve 2).

3. Autoregulation

It is well-known that most organs can locally adjust their resistance to maintain an appropriate blood flow. Against a high perfusing flow the vessels retract, against a failing flow the vessels dilate. The overall sum of these local mechanisms was described by a relationship relating the total vasculature (Vas) to the CO. This mechanism is not instantaneous and its time evolution was formulated as of exponential type.

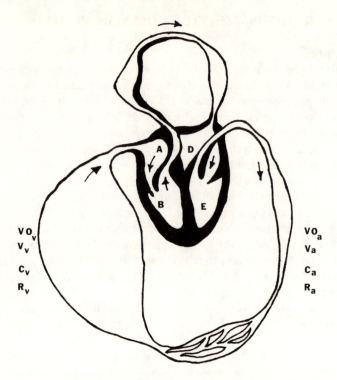

FIGURE 1.   A schema of the circulatory system.

4. The Heart

Heart and lung were integrated together. The capacity of the heart to pump the exact amount of blood it receives from the left atrial was described by the well-known Frank-Starling curve, $CO = f(P_{ra})$ (Figure 2, curve 3). However, as the heart has to pump blood into arteries against an existing pressure, P_a, account was taken for the pressure load on the heart. A factor, APM, as function of P_a, was introduced and the actual cardiac output was defined as $CO = APM \cdot f(P_{ra})$. Finally, another factor, k_4, accounted for an eventual heart failure.

5. The Baro-Chemo System to Stabilize Blood Pressure

Nerve endings, sensitive to wall stretching, have been found in the wall of the great systemic arteries, especially in the carotid-sinus region and in the aortic arch. Neuronal linkage to the brain follows two routes, one via the glossopharyngeal nerves, the other via vagal nerves. Return signals from the brain reach the arterioles, the capacitant vessels, the heart, and the kidney. The above nerve endings are called baro-receptors.

Another type of receptors, the chemo receptors, are sensitive to the changes in the chemical composition of the fluids in certain tissues.

From the viewpoint of blood pressure control, the baro-chemo system acts as a feedback mechanism: a rise in pressure (1) stretches the arterial wall and stimulates the baro-receptors, (2) modifies fluid apport, oxygen and carbon dioxide composition in tissues, and stimulates chemo-receptors. Return signals from the brain dilate the arterioles, decrease heart rate, and lower urinary output. All these effects, in turn, decrease arterial pressure.

Guyton and Coleman formulated the baro-chemo system very roughly. A baro-chemo factor, BC, depending on the pressure level, was split into a baro-component, αBC, and a chemo-component, $(1-\alpha) BC$. The partition was necessary because the baroreceptors "adapt" to sustained pressure: if pressure is kept elevated for several days, the baro-effects degrade to zero. In contrast, the chemo-receptors do not seem to adapt. Recomposition of the adapted baro-component, B(a), and the nonadapted chemo-component gave the "autonomic multiplier", AUM, which was the effector coefficient of the baro-chemo system.

C. The Linearized Equations of Guyton-Coleman Model

As can be seen in the data reported in Section III., averaged arterial pressure covered a range from 85 to 150 mmHg. Between these limits, only the linear portions of the function blocks (Figure 2) were concerned. As a result, we suppose, in this study, that all the function blocks are linear.

The linearized Guyton-Coleman equations can be written as follows:

- $UO = a_1 P_a + b_1$ (pressure-diuresis curve)
- $\dfrac{dEC\ FV}{dt} = I - UO$ (mass balance equation, I = absorption rate)
- $BV = a_2 EC\ FV + b_2$ (partition of fluid volumes)
- $MSP = AUM \cdot (BV - BVO)/(C_a + C_v)$ (definition of MSP)
    $= (a_3 BV + b_3) \cdot AUM$
- $CO = (MSP - P_{ra})/R_{vr}$, where $R_{vr} = R_v + R_a C_a/(C_a + C_v)$ (Guyton's formula)
- $(R_a)_b = k_1/V_{as}$ (definition of structural arterial resistance)
- $TPR = R_a + R_v$ (definition of total resistance)
- $APM = a_6 P_a + b_6$ (load factor of arterial pressure on the heart)
- $P_{ra} = a_5 CO_n + b_5$ (Frank-Starling law, CO_n = intrinsic flow rate of the heart)
- $CO = CO_n \cdot APM \cdot AUM/k_4$ (actual cardiac output)
- $BC = a_7 P_a + b_7$ (stimulation factor for the baro-chemo system)
- $\dfrac{dZ}{dt} = -k_3 Z + \alpha k_3 (1 - BC)$ ($Z = B(a) - \alpha BC$, k_3 = rate constant of the adaptation of

    the baro-receptors)
- $\dfrac{dV_{as}}{dt} = -k_2 V_{as} + a_4 CO + b_4$ (modification of vasculature to regulate flow, k_2 = rate

    constant of the process)

D. Stability and Steady State of Guyton-Coleman Model

The whole set of Guyton-Coleman equations can be synthetized into a system of three different equations, with ECFV(t), Z(t) and V_{as}(t) as unknown functions. The second members cannot be explicitly expressed as functions of ECFV, Z, and V_{as}. However, they can be determined by an approximation procedure.

Although it has been recognized that the system is stable "around normal conditions", the general problem of stability has not been discussed mathematically. Two problems are still unsolved: (1) define the value-domain for the coefficients such that the system is stable and (2) for each set of coefficients, define the value-domain of initial conditions for which the system is stable. The study of these problems may help to understand the reason why, in some specific situations (for example in severe bleeding) some feedback mechanisms completely fail.

Suppose now that the system is stable and that the input to the system is constant. Then the steady state of the system can easily be defined. One obtains, for example:

$$AP(t \to \infty) = A\infty = (I - b_1)/a_1 \qquad (1)$$

This equation shows that hypertension cannot occur without an abnormal (salt or water) absorption and/or an impairment of the renal function.

Steady state values of the other variables can be solved as function of the coefficients as follows:

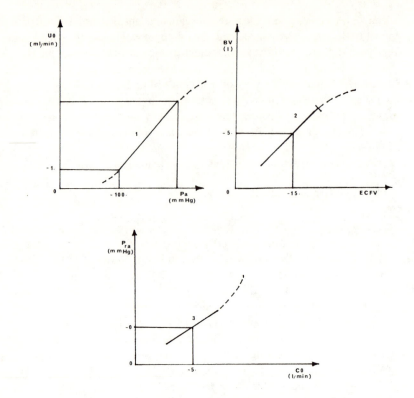

FIGURE 2.   Three function blocks of the Guyton-Coleman model.

$$AUM_\infty = AUM_\infty(I, a_1, b_1, a_7, b_7) \tag{2}$$

$$(V_{as})_\infty = (V_{as})_\infty (I, a_1, b_1, a_7, b_7, a_4, b_4, k_1, k_2) \tag{3}$$

$$TPR_\infty = TPR_\infty (I, a_1, b_1, a_7, b_7, a_4, b_4, k_1, k_2) \tag{4}$$

$$CO_\infty = CO_\infty (I, a_1, b_1, a_7, b_7, a_4, b_4, k_1, k_2) \tag{5}$$

$$BV_\infty = BV_\infty (I, a_1, b_1, a_7, b_7, VRES, k_4, a_4, b_4, k_1, k_2,$$
$$a_3, b_3, a_5, b_5, a_6, b_6, \beta, \gamma) \tag{6}$$

In the above equations, all the coefficients on which the steady states depend were listed. Expressions for the second members are not simple and were not reproduced.

E. Application to Man, Normal Steady State, and Normal Coefficients

To apply the model to man, the coefficients were estimated in such a way that the model predicts the same steady states as observed in man.

Table 1 depicts a set of normal levels (mean-values calculated from the data on the first group of 40 normotensives, see Section III following). Normal values for the coefficients, obtained by Guyton and Coleman, have been slightly modified to account for the present data. The results are depicted also in Table 1.

F. Short-Term Behavior of the Model

Short-term behaviors of the model have been extensively investigated by Guyton and collaborators: absorption and/or internal functions of the system were artificially changed, the resulting theoretical solutions calculated and confronted with observations in dogs. One celebrated experiment[8] was conducted as follows. Two thirds of the renal mass was removed and absorption of water and salt was doubled in a series of six dogs. Heart rate, blood pressure, blood volume, and cardiac output were followed for 2 weeks. In terms of the model, one looked for the solution starting from the normal steady state, after that, the renal function curve had been resettled to the right and the absorption coefficient doubled.

The model predicts a two-phase hemodynamic change. In the first phase, arterial pressure increased, due mainly to renal mass reduction, TPR decreased under autonomic reflex control, cardiac output increased because of over infusion and reduced excretion that increased blood volume. In the second phase CO decreased rapidly because of tissue local autoregulation while pressure and resistance remained elevated. Hypertension has been created, with one almost normal blood flow. The experimental results matched perfectly with the model prediction. They were powerful arguments in favor of Guyton's model (details of this experimentation were reported in Reference 8).

III. ANALYSIS OF HUMAN DATA IN ESSENTIAL HYPERTENSION

This section discusses a set of clinical data in essential hypertension.[4,5] To prepare the matching of observations with model predictions, the data were "smoothed" by the sliding-mean method. Details of the method are described below.

A. Patients

The data concerned 196 male normotensive and hypertensive adults. To minimize the role of age in the study, only individuals aged 20 to 40 years were included.

The patients were either untreated or had discontinued therapy for at least 4 weeks before the study. During hospitalization for 6 days, dietary sodium intake was 110 mmeg/day. A steady state of sodium balance was established on the basis of body weight, sodium intake, and urinary output.

Extensive investigations were performed to detect secondary hypertension: blood and urinary electrolytes, and catecholamine determinations, endogenous creatinine clearance, time intravenous urography and/or renal asteriography. All the patients included in the study were listed as having essential hypertension. Optic fundi were normal. No subject had any cardiac or neurologic involvement.

B. Measurements

After overnight fasting, the patients were brought to the hemodynamic laboratory without premedication. A thin-wall needle was inserted into a brachial artery to measure blood pressure. Cardiac output was determined by injecting a colored liquid (indocyanine green) in a vein and measuring its appearance in an artery. Blood volume was assessed by injecting radio-iodinated albumin in a vein; albumin is distributed only in blood. By studying albumin kinetics, one derived its distribution volume. Details of these methods can be seen in current medical textbooks.

C. Analysis of the Data by the Sliding Mean Method

The standard method to analyze clinical data was to divide the whole population into a "normotensive" group and a "hypertensive" group, and to compare mean values of (and correlation between) the variables in the two groups. This method is questionable because hypertension is a progressive pathology; there is no clear-cut frontier between normalcy and disease.

To get an insight into the evolution of the pathology, one possibility might be to divide the population observed into different groups with different pressure ranges. One investigates the changes of the hemodynamic pattern in these groups following the pressure rise. A more suggestive method would be to classify the subjects by increasing blood pressure, to form overlapping subgroups of subjects "moving" in the direction of high pressure and to examine hemodynamic changes parallel to the pressure rise. This is the sliding-mean method detailed below.

Consider a population of N individuals, classified by increasing arterial blood pressure and indexed by a figure i, i = 1,2,..N. We form subgroups of n subjects from the total population as follows:

subgroup 1 = individuals 1, 2, ..., n
subgroup 2 = individuals 2, 3, ..., n + 1
... ...
subgroup N − n + 1 = individuals N − n + 1, ..., N

In each group, we determine mean values of the hemodynamic variables and correlations between them. The results describe the main hemodynamic pattern of the subgroup. As two adjacent subgroups differ only by one individual from one group to the next, mean values, correlation coefficients, and regression coefficients change very slowly. Such variations depict a smooth picture of the hemodynamic pattern accompanying the increase in blood pressure. The picture might disclose the pressure levels where a certain variable or correlation is more definitely affected.

D. Results

From the 196 patients classified by increasing arterial pressure, we formed 157 subgroups of 40 patients, each subgroup differing from the following by one individual. Figure 3 depicts from subgroup 1 to subgroup 157, mean values of heart rate ($\overline{HR}$), blood volume ($\overline{BV}$), and cardiac output ($\overline{CO}$) against mean value of arterial pressure ($\overline{AP}$). Figure 3 shows that heart rate sharply increases for $\overline{MAP}$ between 85 mmHg and 95 mmHg, and remains elevated at higher pressures. Cardiac output has an increasing phase for pressure between 85 mmHg and 110 mmHg, and a decreasing phase for pressure between 130 mmHg and 147 mmHg. Over the whole range of pressure, blood volume remains unchanged.

Figure 4 depicts the correlation coefficients of the cardiac output-heart rate and of the cardiac output-blood volume relationships against mean values of arterial pressure. Figure 4 shows that the correlation CO-HR is significant only in the normotensive ranges, while the correlation CO-BV is significant only in the hypertensive ranges.

E. Statistical Inference of the Results

To ascertain statistical meaning to the above geometric findings, classical tests were performed in *distinct* subgroups isolated from the global population.

1. Mean Values

From the results of Figure 3, the population was divided into two groups, A and B, to test the increase of heart rate from normotensive to hypertensive patients, and into three groups, C, D, and E, to test the changes in cardiac output. Table 2 shows that heart rate was significantly increased from Group A to Group B. Table 3 gives the mean values of heart rate CO and BV in Groups C, D, and E. Table 4 depicts the results of a variance analysis performed to test the equality of mean cardiac output in Groups C, D, and E. As the calculated F-value was higher than the F-value read from the statistical tables, mean values of cardiac output were not the same in the three groups. One concludes that cardiac output among the three groups was highest in Group D.

2. Correlations

To show the effectiveness of the changes in correlations, we considered the following groups:

Group F = patients numbered 1 to 93
Group G = patients numbered 134 to 196

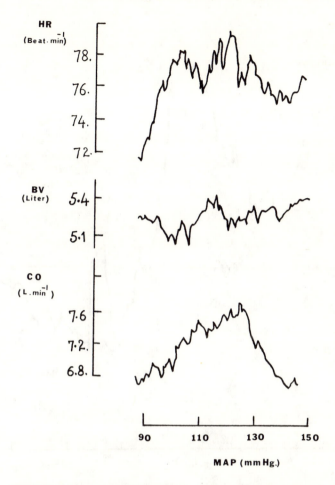

FIGURE 3.   Mean values of heart rate, blood volume, and cardiac output
depicted against mean values of mean arterial pressure. The successive
points were joined linearly.

Table 5 shows that the correlation CO-HR was significant in Group F but not in Group
G. In contrast, the correlation cardiac output-blood volume was significant in Group G but
not in Group F. Furthermore, from Group F to Group G, the two correlations were significantly changed ($p < 0.05$).

IV. GUYTON-COLEMAN MODEL AND ESSENTIAL HYPERTENSION

In this section an attempt is made to discuss the results of Section III. in the framework
of the Guyton-Coleman model. Clinical data are identified with steady state levels in the
model. Observed heart rate is identified with the model's autonomic multiplier, after multiplication by a constant factor, matching the normal heart rate to normal AUM ($= 1$). The
latter identification was proposed by Guyton et al.[8]

The problem under discussion is to look for the possible impairments (changes of coefficients of the model from their normal values) to account for the results in Figures 3 and
4. We proceeded in two steps: a ''sensitivity matrix'' was first constructed to detail the
contribution of each coefficient of the system to an abnormal steady state. The sensitivity
matrix was then used as a basis for the discussion.

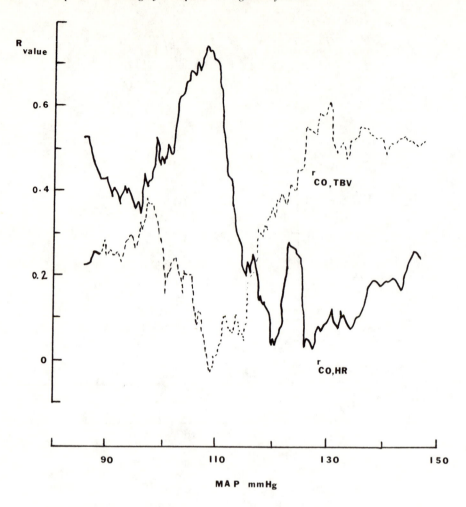

FIGURE 4.   Correlation coefficients of the cardiac output-heart rate relationship and of the cardiac output-blood volume relationship depicted against mean values of mean arterial pressure. The successive points were joined linearly.

A. Abnormal Steady States, Contribution of Each Coefficient to an Abnormal Steady State: the Sensitivity Matrix

Suppose that the Guyton-Coleman is in an abnormal steady state, for example with Pa = 140 mmHg and CO = 8 ℓ. From Equations 1 to 6 in Section II.D., one concludes that at least one of the coefficients I, a_1, b_1, a_7, b_7, a_4, b_4, k_1, and k_2 is abnormal. However such information is only qualitative. To evaluate the quantitative contribution of each coefficient to an abnormal steady state, we proceed as follows.

Consider a particular coefficient, for example b_1, and a particular variable, for example CO. Suppose that b_1 is doubled from its normal value (that is, from -4.4 to -8.8). Then the percentage of change of cardiac output from its normal value is

$$100 \times \frac{CO\ (I, a_1, 2b_1, a_7, \ldots) - CO\ (I, a_1, b_1, a_7, \ldots)}{CO\ (I, a_1, b_1, a_7, \ldots)} = 5.76$$

The fact that b_1 is doubled means that the UO-AP renal curve is reset to the right (Figure 5). As a consequence, at the same pressure $\overline{OP}$, the urinary output is reduced from $\overline{OA}$ to $\overline{OA'}$. This modification might account for a reduction of the renal mass. An analog calculation of the total peripheral resistance gives:

Table 2
TEST FOR THE INCREASE OF HEART RATE FROM NORMOTENSIVES TO HYPERTENSIVE PATIENTS

	Group A	Group B
Patient no.	1 to 40	41 to 196
Mean arterial pressure range (mmHg)	63 to 95	96 to 174
Heart rate (beat/min)	72 ± 2	77 ± 1[a]
Cardiac output (ℓ/min)	6.79 ± 0.24	7.22 ± 0.12[b]
Blood volume (ℓ)	5.30 ± 0.12	5.29 ± 0.06[b]

[a] $p < 0.05$.
[b] p; Not significant.

Table 3
MEAN VALUES OF HEART RATE, CARDIAC OUTPUT, AND BLOOD VOLUME IN GROUPS C, D, AND E

	Subgroup		
	C	D	E
Patient no.	1—93	94—150	151—196
Mean arterial pressure range (mmHg)	63—110	111—130	131—174
Heart rate (beat/min)	75 ± 1	78 ± 2	77 ± 2
Cardiac output (ℓ/min)	6.99 ± 0.15	7.54 ± 0.20	6.87 ± 0.26
Blood volume (ℓ)	5.22 ± 0.07	5.34 ± 0.10	5.37 ± 0.13

Table 4
VARIANCE ANALYSIS FOR CARDIAC OUTPUT IN GROUPS C, D, AND E

Origin of variations	Sum of squares of deviations	Variance	Degree of freedom	Calculated F-value
Inter-groups	14.71	7.36	2	3.10
Residual	457.70	455.32	193	
Total	472.41		195	

$$100 \times \frac{\text{TPR} (I, a_1, 2b_1, a_7, \ldots) - \text{TPR} (I, a_1, b_1, a_7, \ldots)}{\text{TPR} (I, a_1, b_1, a_7, \ldots)} = 72.4$$

One concludes that the above impairment of the renal function leads the system to a new steady state with higher cardiac output (+ 6%) and higher resistance (+ 72.4%). The effect of the impairment is $72.4/5.76 = 12.6$ more important on the levels of resistance than on those of cardiac output.

The calculations can be performed on all the other variables. Table 6 (the sensitivity matrix) depicts the percent of changes from normal values of seven variables, when each of the coefficients is doubled from its normal value. It is seen from Table 6 that:

1. A change of the renal function (coefficients a_1, b_1) has tremendous effects on Pa, TPR and vasculature (Vas), but relatively few effects on fluid volumes, CO and autonomic control. The coefficients a_1 and b_1 act on pressure, volumes, and cardiac output *in the same direction*.

Table 5
CORRELATION COEFFICIENTS (r)
BETWEEN CARDIAC OUTPUT, HEART
RATE, AND BLOOD VOLUME IN TWO
DISTINCT GROUPS, F AND G

	Group F	Group G
Patient no.	1—93	134—196
Mean arterial pressure range (mmHg)	63—110	121—174
r (CO-HR)	0.56[a]	0.28
r (CO-BV)	0.18	0.50[a]

Note: See Table 1 for abbreviations.

[a] $p < 0.001$.

2. A change of the autonomic system (coefficients a_7, b_7) has relatively few effects on volume, resistance, and cardiac output.
3. A change in the rate of creation of vasculature (coefficients a_4, b_4), in the rate of destruction of vasculature (coefficients k_2) or in the resistivity of the vascular bed (coefficient k_1) has comparable effects, *but in the opposite direction* on TPR and CO. Their effects on fluid volumes are comparatively few.
4. Blood volume depends on 19 coefficients of the system. However, except for two groups, doubling each of the coefficients changes blood volume only 5 to 10%. The two exceptional groups are the coefficients a_2, b_2, which define the partition of blood and extracellular fluid volumes, and the coefficients a_3, b_3, which characterize the capacitance of the vascular bed.
5. The coefficients α (partition of the autonomic control into baro- and chemo components) and k_3 (adaptation of the baro-chemo system) have no effect on the steady state of the system.

Table 6 is of fundamental importance in the following discussions.

B. Study of Mean Values

To account for the observed pattern of Figure 3, we discussed the possible causes for the elevation of pressure, heart rate, and cardiac output at moderate pressure, and the reasons for the decrease of cardiac output at elevated pressure.

1. Elevation of Pressure

From Table 6 it was seen that steady state pressure increases only if the absorption rate increases and/or the renal function is impaired. Figure 6a depicts the first sequences of simulation in which absorption was supposed normal but both a_1 and b_1 were supposed decreased. The final values of a_1 and b_1 were calculated to get a pressure of 150 mmHg. The whole variations of a_1 and b_1 were then divided into 100 steps. In each step, steady state values of Pa, CO, AUM, and BV were calculated. Finally, steady state values of AUM, BV, and CO were depicted against those of Pa. The successive points were joined linearly.

The simulated enfeeblement of the renal function resulted in a potent increase of mean arterial pressure, a decrease of autonomic control, and an increase of blood volume and total peripheral resistance. In contrast, cardiac output remained almost constant.

Except for the rise in pressure, Figure 6a shows that an impairment of the renal function alone *could not* account for the data in Figure 3.

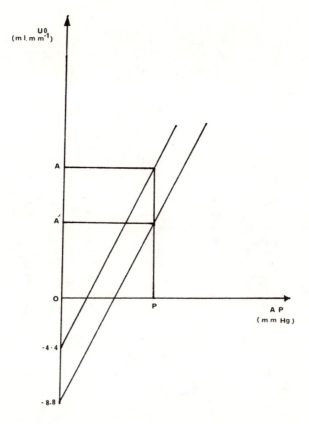

FIGURE 5.   Two urinary output-arterial pressure curves with intercepts
$b_2 = -4.4$ and $b_2 = -8.8$.

2. Increase of Heart Rate in Mild Hypertension

Table 6 shows that a renal impairment decreases heart rate. To account for the observed increase in heart rate in moderate hypertension, the only possibility was to suppose that a_7 and/or b_7 potently increased at these pressures.

3. Increase of CO at Moderate Pressure

Table 6 shows that the above two impairments (kidney and baro-receptors) only slightly modified the level of CO. To account for the rise in CO at moderate pressures, we supposed that the coefficients a_4, b_4 were impaired at these pressures. Figure 6b depicts a second set of simulations in which the renal function (coefficients a_1, b_1), the autonomic nervous control (coefficients a_7, b_7), and the autoregulation of blood flow (coefficients a_4, b_4) were modified. Figure 6b shows that AUM increased and remained elevated. (These variations accounted for the observed heart rate.) However, CO sharply increased while TPR only mildly increased.

4. Decrease in Cardiac Output at High Pressure

To account for the decrease in CO and an increase in TPR in established hypertension, we suggested that at these pressure ranges the resistivity of the vascular bed was impaired. This modified coefficients k_1 and k_2. Figure 6c depicts the final simulations. The predicted curves (Figure 6c) were quite comparable with observations (Figure 3).

C. Study of Correlations

At moderate pressure, CO and AUM varied in the same direction. At high pressures, CO and BV varied in the same direction. These facts explained ''qualitatively'' the results of

Table 6
PERCENT OF CHANGES OF THE VARIABLES FROM NORMAL STEADY STATES WHEN EACH OF THE COEFFICIENTS IS DOUBLED FROM NORMAL VALUE

Coefficient ECFV	ECFV BV[a]	Vas	TPR	Pa	CO	AUM
I	2.3	−18.7	16.3	18.5	1.9	−2.1
a_1	−5.9	126.9	−42.7	−50	−12.8	5.6
b_1	9.5	−52.4	72.4	81.5	5.3	−9.2
a_7	2.8	−10.1	−1	0	1.0	−11.2
b_7	−5.2	31	3.2	0	−3.1	36.2
a_4	−6.8	−53	89.9	0	−47.3	0
b_4	13.5	155.3	−48.6	0	94.5	0
k_2	−1.1	−9.9	8.8	0	−8.1	0
k_1	−1.1	80.2	8.8	0	−8.1	0
VRES	7.6	27.6	2.8	0	−2.8	0
a_3	−50.0	0	0	0	0	0
b_3	−86.0	0	0	0	0	0
a_6	1.3	0	0	0	0	0
b_6	−3.3	0	0	0	0	0
a_5	6.1	0	0	0	0	0
b_5	−6.1	0	0	0	0	0
k_4	6.1	0	0	0	0	0
β	9.4	0	0	0	0	0
γ	4.6	0	0	0	0	0
a_2	50 0	0	0	0	0	0
b_2	−19 0	0	0	0	0	0
α	0	0	0	0	0	0
k_3	0	0	0	0	0	0

Note: b_2 is changed from 0 to 1.

[a]   Values for the ECFV and BV columns are identical except for lines a_2, b_2. See Table 1 for abbreviations and units.

the correlation study (Section III.). In fact the interpretation of the existence and the change of a correlation performed on steady state values is a subtle problem. A forthcoming report will discuss this problem in detail.[9]

D. Impairments

Variation of the coefficients accounting for the pattern of Figure 6c are shown in Figure 7. By displacing a vertical rule at any Pa level (between 90 and 150 mmHg, Figure 7) one may determine, at the intersections of the rule with the curves, the values of the coefficients contributing to define this pressure level. By the same method in Figure 6c, one may read the corresponding values of AUM, BV, CO, and TPR.

Note that in Figures 6 and 7, arterial pressure has been used on the abscissa for the purpose of presentation. This did not mean that, in these figures, Pa was an independent variable and the other variables or coefficients represented in ordinates were dependent variables.*

V. CONCLUSION AND DISCUSSION

This chapter presents a new approach to clinical data based on animal models. The approach is illustrated by the application of the 1967-Guyton-Coleman circulatory model to essential hypertension.

* In these simulations, we implicitly supposed that the impairments of the coefficients were progressive and *monotonous*.

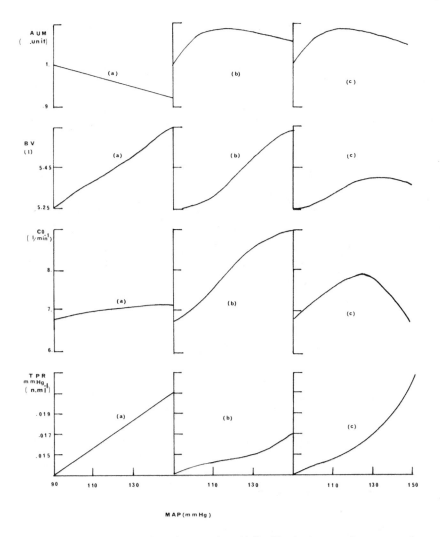

FIGURE 6.   Steady state values of autonomic multiplier, blood volume, cardiac output, and total peripheral resistance depicted against steady state values of arterial pressure when the renal function was progressively impaired (curves a); when the renal function, the autonomic control, and the vasculature were progressively impaired (curves b); and when the renal function, the autonomic control, the vasculature, and the systemic resistivity coefficient were progressively impaired (curves c).

1.   The starting point of the article was the following fundamental remark: in most animal studies of hypertension, hemodynamics, artificially and acutely impaired, were observed as functions of time. In contrast, in most human studies, only steady states of hemodynamics were determined. Clinical data, in particular, are steady state levels and must be compared with steady states in animal models.

2.   The 1967-Guyton-Coleman model has been used to interpret our data. Of course, a simpler model might be sufficient for the study of cardiac output, heart rate, and blood volume. However, the Guyton-Coleman model was well-known, has been repeatedly tested, and included the most important mechanisms of blood pressure control. Based on this model, Table 4, which details the effects of different impairments on the global steady state, might be useful for other studies, independent of the data in this article. When extended to include the renin-angiotension-aldosterone system, weight, and age, this model might be most useful to interpret the clinical data on hypertension available in the literature.

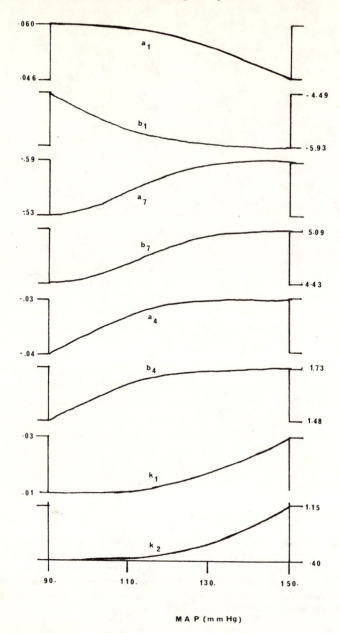

FIGURE 7.   Variation of the regulation coefficients accounting for the pattern of Figure 6c depicted against steady state of arterial pressure.

3.   Human data can only be analyzed by statistical methods. A geometric screening technique has been proposed to explore the data and to point out the most important changes of hemodynamics from normotensive to hypertensive patients. The underlying statistical basis for this technique is the linear regression and correlation analysis. The problem is to detect and localize with a confidence interval, the points of the regression curve *where the curvature radius is the smallest* (Figure 8). To the authors' knowledge, the problem is not yet solved.

4.   Figure 6c was obtained by a careful study of Table 4. This figure depicts a possible schema of multiple impairments leading to the multiparametric observations of Figure

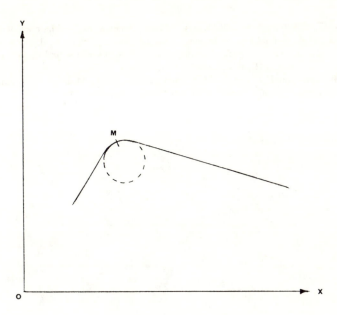

FIGURE 8.   Localization of the point of smallest curvature radius on a
regression curve: an open problem.

3. Figure 6a and b, or other analog simulations, might be useful for eventual animal
experimentations, in which one might determine the steady states of hemodynamics
resulting from progressive impairments of some control loop.

In conclusion, by considering the application of the Guyton-Coleman model to essential
hypertension, we wanted to detail a new approach to clinical data by the use of models.

This approach was an example of the application of computer modeling in physiological
research.

REFERENCES

1. **Amery, A., Fagard, R., Lijnen, P., and Straessen, E. D.,** *Hypertensive Cardiovascular Disease: Path-ophysiology and Treatment,* Martinus Nijhoff Publ., The Netherlands, 1982, chap. 5.
2. **Guyton, A. C., Jones, C. E., and Coleman, T. G.,** *Circulatory Physiology: Cardiac Output and its Regulation,* W. B. Saunders, Philadelphia, 1973.
3. **Guyton, A. C.,** *Arterial Pressure and Hypertension,* W. B. Saunders, Philadelphia, 1980.
4. **Safar, M. E., Chau, N. P., London, G. M., Weiss, Y. A., and Milliez, P. L.,** Cardiac output control in essential hypertension, *Am. J. Cardiol.,* 38, 332, 1976.
5. **Chau, N. P., Safar, M. E., Weiss, Y. A., London, G. M., Simon, A. Ch., and Milliez, P. L.,** Relationships between cardiac output, heart rate and blood volume in essential hypertension, *Clin. Sci. Mol. Med.,* 54, 175, 1978.
6. **Chau, N. P., Safar, M. E., London, G. M., and Weiss, Y. A.,** Essential hypertension: an approach to clinical data by the use of models, *Hypertension,* 1, 86, 1979.
7. **Guyton, A. C. and Coleman, T. G.,** Long-term regulation of the circulation: interrelationships with body fluid volumes, in *Physical Bases of Circulatory Transport: Regulation and Exchange,* edited by Reeve, E. B. and Guyton, A. C., Eds., W. B. Saunders, Philadelphia, 1967, 179.

8. **Guyton, A. C., Coleman, T. G., Cowley, A. W., Jr., Norman, R. A., Manning, R. D., and Liard, J. F.,** Relationships of fluid and electrolytes to arterial pressure control and hypertension: quantitative analysis of an infinite gain feed-back system, in *Hypertension Mechanism and Management,* Onesti, G., Kim, K. E., and Moyer, J. H., Eds., Grune & Stratton, New York, 1973, 25.
9. **Chau, N. P., Coleman, T. G., Safar, M. E., and London, G. M.,** Meaning of the cardiac output blood volume relationship in essential hypertension, *Am. J. Physiol.,* in press, 1983.

Chapter 7

COMPUTER MODELING IN FOOD PROCESS ENGINEERING SYSTEMS

Pedro Francisco Castillo Monteza and Ramu M. Rao

TABLE OF CONTENTS

I. INTRODUCTION

Food can be considered as a biological system containing or consisting of carbohydrates, fats, proteins, and other supplementary substances such as minerals used in the body of an organism to sustain growth, repair, and vital processes and to furnish energy. From the place of origin to the point of consumption, more than 95% of our food is processed using such methods as canning or freezing. Food processing is essential if a population is to be fed. In most cases food processing causes a reduction in the nutritional value of a food. The reduction of nutrient retention in the processed foods has been a major concern of food manufacturers and this concern has constantly prompted studies into ways to reduce quality destruction in the heat sterilization process and different means to predict these losses according to the conditions of the process.

Rapid developments in the computer industry have brought about the availability of a wide variety of low-cost microprocessors and programmable controllers along with related peripheral equipment that have advanced the state-of-the-art in computer-based process control technology to a considerable extent. Although the field of food technology reaps some benefits of computerization now, the continued influx of computers for use in information management, statistical analysis and modeling, instrumentation, process control and data processing systems is projected to increase dramatically due to cost reduction, increased availability, and expanded sophistication in basic research.

The formulation of mathematical models for the prediction of the loss of nutrients has been tried in the past utilizing different approaches.[5,19,26,42,44,45]

Most recently, Barreiro et al.[6] and Lenz and Lund[33] presented theoretical models for the prediction of the loss of nutrients during thermal processing of foods. Lenz and Lund[33] developed the lethality Fourier number method for calculating the center point sterilizing value of a thermal process applied to conduction-heating food by combining first-order destruction kinetics with the Arrhenius equation and dimensionalizing this equation.

Barreiro et al.[6] developed their model by solving analytically the heat conduction equation for a finite cylinder under nonstationary conditions.

In another study, Herrera[25] applied a similar approach for the prediction of nutrient retention in foods heated by conduction in rectangular cans.

Teixeira[45] used the model proposed in his earlier work[44] to simulate the thermal processing of a canned food product in order to predict the effects on the level of thiamin produced by various surface temperature policies of equal sterilizing value. He concluded that variations in container geometry which permit a more rapid heating of the product can be very effective in achieving a significant improvement in the nutritional value of thermally processed foods. These conclusions support the use of flexible packaging in thermal processing.

Retort pouches are flexible packaging containers developed in the U.S. in accordance with military requirements. They were developed by Continental Can Corp. with the purpose of packing foods for the space program. The information accumulated by Continental was utilized by Toyo Seikan Co., Japan,[28] and, in 1968, the RP-F pouch was developed.

The retort pouches are classified in terms of their performance properties in two main categories: those which use aluminum foil and those which do not use aluminum foil.[37] Since the former category uses aluminum foil, the food contents have an outstanding preservability on account of the complete interception of oxygen and light rays. The latter type is transparent and has the advantage that the foodstuffs contained inside are clearly visible.[49]

The retort pouch (RP-F), which is called "can in pouch form", is composed of three layers: a polyester film, aluminum foil, and polyolefin system film. Each of the layers are firmly bonded together by means of a heat-resistant bonding agent.

The advantages and characteristic features of RP-F are described as follows:[2,3,13,28,46] (1) heat sterilization by means of steam at a temperature of 250°F is possible; (2) the heat transfer with respect to the food contents is faster than in the case of cans with an identical volume; (3) gases such as air and oxygen, as well as light rays and moisture, can be shut out completely; (4) all of the component materials which are used, in particular the polyolefin system film on the inside surface, are safe from the standpoint of food sanitation; (5) they have good heat and hermetic sealing properties; (6) they are convenient

In order to demonstrate as to how computer modeling can be applied to a biological system such as food processing, a computer model which can predict the retention of nutrients in foods packaged in retortable pouches and sterilized by thermal conduction is presented in this chapter. The model was validated experimentally by determining the fraction of thiamin retained in a simulated food, packaged in a retortable pouch, and processed thermally in a still retort.

II. REVIEW OF LITERATURE

The effect of the heat sterilization process for canned foods on the quality and nutrient retention of the food has been a major concern of food processors since the beginning of the canning industry. This concern has constantly prompted study into ways to reduce quality destruction in the sterilization process and different ways to predict these losses according to the conditions of processing.

Since Stumbo[40] presented certain bacteriological considerations for the purpose of demonstrating that in process evaluation procedures greater attention should be given to the kinetics of microorganism inactivation when subjected to moist heat and to the mechanism of heat transfer within the food container during process, a number of graphical and mathematical procedures have been proposed.[12,41,42] These procedures have been established mainly for estimating the effectiveness of thermal processes in terms of their capacities to reduce the number of viable microorganisms in containers of food for which the heating characteristics were known.

These procedures were developed for the estimation of microbial destruction and not for estimating vitamin, color, and enzyme degradation even though many of these food constituents, and perhaps all, degrade in accordance with first order kinetics.

Ball and Olson,[5] Teixeira et al.,[44] and Hayakawa[21] presented methods for estimating nutrient degradation in canned foods during thermal processing.

The models developed by Stumbo[42] and Ball and Olson[5] were based on mass average sterilizing value (F_s). The model proposed by Stumbo[42] may be used only for small values of D_o and Z_o, and the results are only approximations to the values obtained experimentally. The method of Ball and Olson[5] requires major computational effort. Beyond this, because of the finite layer procedure used in integrating lethal effects, and because of inadequate accounting for cooling curve lags, high accuracy in estimations would not be expected.

Hayakawa[21] developed a formula for calculating the mass average sterilizing value (F_s) by using all non-negligible terms in the formula for heat conduction in canned foods during heat processing. From the derived formula, tables of numerical values for a dimensionless group containing F_s were obtained. This model also requires experimental heat penetration data in order to obtain the values of f_h and j which are necessary for the calculation of one of the dimensionless groups. The dimensionless group method, though it may give acceptable accuracy in estimations, requires major computational effort for calculating and interpolating values for five dimensionless groups.

Different models for estimating the central and transient temperatures of canned foods during heat processing have been discussed in the literature by Hayakawa and Ball,[14,15] Hayakawa,[17-19,22] and Hayakawa and Timbers.[16]

A completely theoretical model was proposed by Teixeira et al.[44] who developed a numerical computer model to simulate the thermal processing of canned foods that would simultaneously predict the lethal effect of a heat process on bacteria and its destructive effect on thiamin. Teixiera[45] used this model to simulate the thermal processing of a canned food product in order to predict the effects on the level of thiamin retention produced by various surface temperature policies of equal sterilizing value. He concluded that variations in container geometry which permit a more rapid heating of the product can be very effective in achieving a significant improvement in the nutritional value of thermally processed foods. These considerations support further development in the use of flexible packaging in thermal processing.

Jen et al.[26] modified the original Stumbo method to obtain one that would apply equally well to both thiamin and bacterial inactivation, or to the inactivation of any other thermally vulnerable factor exhibiting first order kinetics. In devising the method, a new equation was derived for integrating heat effects throughout the food in the container during heating and cooling. In order to implement this method, it was necessary to extend f_h/U:g relationship tables for Z values in the higher range. The new tables cover Z values (for every other Z value) from $Z = 8$ to $Z = 80$.

The results obtained with this manual method compare favorably with results obtained by other more complicated methods for the same purpose, namely those of Ball and Olson,[5] Teixeira et al., and Hayakawa.[19]

The accuracy and validity of Stumbo's modified mathematical method and computer model of Teixeira et al.[44] were studied by Teixeira et al.[75] The experimental evidence obtained showed agreement with the calculated values, in all cases, well within the $\pm 6\%$ expected error, and confirms the validity of both the computer model developed by Teixeira et al.[44] and Stumbo's method as modified by Jen et al.[26] A similar confirmatory comparison was reported by Jen et al.[26] which included the assay determinations reported by Hayakawa[19] and predictions of percent thiamin retention by the methods of Ball and Olson,[3] Hayakawa,[21] Teixeira et al.,[44] and Jen et al.[26]

Lenz and Lund[33] developed the lethality-Fourier number method for calculating the center point sterilizing value of a thermal process applied to conduction-heating foods by combining first order destruction kinetics with the Arrhenius equation and dimensionalizing this equation.

The lethality-Fourier number method was extended by the same authors, Lenz and Lund[31,33] so that mass-average retention of heat-labile quality factors (e.g., color and nutrients) could be calculated for conduction heating foods. The method was evaluated by comparing predicted mass-average retention of thiamin and chlorophyll in food purees to those obtained experimentally.

The method was also compared to several other methods which have been developed for calculating mass-average retention of heat-labile components with results at least as accurate as other available techniques. Lenz and Lund[33] also investigated the effect of biological variability in destruction rates of microorganisms and quality factors and in thermal diffusivity on accuracy of thermal process calculations.

Another theoretical model for the prediction of the loss of nutrients during thermal processing was developed by Barreiro et al.[6] This model was developed by solving analytically the heat conduction equation for a finite cylinder under nonstationary conditions. This model has the advantage that it may be applied without the restriction related to the final temperature of cooling of Teixeira's model in which the f_h/U:g water and the final temperature of the product.

Barreiro et al.[7] utilized the model they developed in order to optimize the thermal processing which will result in the maximization of nutrient retention.

The model developed by Barreiro et al.[6] was adapted by Herrera[25] for the prediction of the loss of nutrients in rectangular cans with results within 90% confidence intervals of the experimental values.

Several other models have been developed for the prediction of the stability of different nutrients in foods. These models take into consideration several factors which could affect the rate of degradation including time, temperature, moisture content, humidity, oxygen content, and pH. These models are well documented in the literature.[27,30,38,47,48]

III. DEVELOPMENT OF THE MODEL

A theoretical model for the prediction of the degradation of nutrients which follow first order kinetics degradation in foods heated by conduction in retortable pouches is developed in this chapter.

In order to obtain the model, the retortable pouch is approximated to a rectangular parallelepiped which permits the application of the heat conduction equation[8,29] for the prediction of the temperature at any point of the container as a function of time under nonstationary conditions.

This equation is solved analytically in rectangular coordinates for both the heating and the cooling periods of thermal processing. The two equations obtained are included in the equation describing first order kinetics.[10]

In order to include these temperatures into the equation, the reaction velocity constant, k, is substituted by the natural logarithm of 10 over D, in which D is the time required to destroy 90% of the nutrient at a temperature, $T_i(x,y,z)$.[10]

After the proper substitutions are made, the differential equation obtained is integrated between the initial concentration of the nutrient and the remaining fraction of nutrients at any point of the container of coordinates x, y, and z at the end of the processing period. The result of this integration gives the fraction of nutrient retained at one point of the container of coordinates x, y, and z at the end of processing.

Finally, the function obtained is averaged for the complete volume of the pouch about its rectangular dimensions to obtain the average fraction of nutrient retained at all points of the pouch.

A. Limitations

The following limitations or assumptions should be taken into consideration for the development of the model:

1. Heat is transferred by conduction in the product inside the pouch and the thermal diffusivity of the product is constant during the heating and cooling periods.
2. The product is packed in retortable pouches whose dimensions are approximated to a rectangular parallelepiped of dimensions a, b, and c.
3. Heat is transferred through all sides of the pouch.
4. The heat resistances of the three-ply components of the pouch are negligible compared to the thermal resistance of the product.
5. The surface temperature of the product is constant and equal to the temperature of the heating medium.
6. The initial temperature of the product, T_o, and the initial concentration of the nutrient in the food, C_o, are constant and uniform throughout the product.
7. The temperature at the beginning of the cooling period is the same as the temperature at the end of the heating period which is distributed as a function, f(x,y,z).
8. The degradation of the nutrient follows first order kinetics and the reaction velocity constant, k, follows the Arrhenius equation.
9. The nutrient degradation during the lag occurring at the beginning of the heating and cooling periods is negligible.

B. Temperature Prediction During the Heating Period

The temperature at any point of the rectangular pouch for a time, t, less than or equal to the total heating time ($t \leq fht$) is obtained by solving the heat conduction equation proposed by Fourier[9,23] for the temperature distribution in a rectangular parallelepiped:

$$\frac{\partial^2 T}{\partial x^2} + \frac{\partial^2 T}{\partial y^2} + \frac{\partial^2 T}{\partial z^2} = \frac{1}{\alpha}\frac{\partial T}{\partial t} \qquad (1)$$

where T = temperature, t = time, and α = thermal diffusivity.
Prediction of $T_1(x)$ during the heating period:

$$\frac{\partial^2 T_1}{\partial x^2} = \frac{1}{\alpha}\frac{\partial T_1}{\partial t}$$

Initial condition:

$$T_1(x,0) = f(x) = 1$$

Boundary condition:

$$\left.\frac{\partial T_1}{\partial x}\right|_{\substack{x\ =\ 0 \\ x\ =\ a}} = T_1\binom{0}{a},t) = 0$$

$$T_1(x,t) = X(x)\theta(t)$$

$$\left.\frac{\partial T_1}{\partial x} = x\ '(x)\theta(t)\right|_{\substack{x\ =\ 0 \\ x\ =\ a}} = T_1(0,t)h = T_1(a,t)h$$

$$\frac{\partial^2 T_1}{\partial x^2} = X\ ''(x)\theta(t)\ ; \ \frac{\partial T_1}{\partial t} = X(x)\theta'(t)$$

$$X''(x)\ \theta(t) = \frac{1}{\alpha}X(x)\theta'(t)$$

Divide both sides by $X(x)\theta(t)$

$$\frac{X''}{X} = \frac{1}{\alpha}\frac{\theta'}{\theta} = -\lambda^2$$

$$\frac{1}{\alpha}\frac{\theta'}{\theta} = -\lambda^2$$

$$\theta' + \alpha\lambda^2\theta = 0$$

$$\theta(t) = Ae^{-\alpha\lambda^2 t}, \ A = \text{constant}$$

$$\frac{X''}{X} = -\lambda^2$$

$$X'' + \lambda^2 x = 0$$

$$X(x) = L\cos\lambda x + M\sin\lambda x$$

$$T_1(x,t) = (L\cos\lambda x + M\sin\lambda x)Ae^{-\alpha\lambda^2 t}$$

$$A = LA; \ B = MA$$

$$T_1(x,t) = (A\cos\lambda x = B\sin\lambda x)e^{-\alpha\lambda^2 t}$$

$$T_1(0,t) = Ae^{-\alpha\lambda^2 t} = 0$$

$$A = 0$$

$$T_1(a,t) = (B\sin\ a)e^{-\alpha\lambda^2 t} = 0$$

$$B \neq 0$$

To satisfy this condition, $\sin \lambda a$ must be zero or

$$\lambda = \frac{n\pi}{a} \text{ where } n = 1,2,3\ldots$$

Therefore, there exists a different solution for each integer, n, and each solution has a separate integration constant B_n. Summing these solutions:

$$T_1(x,t) = \sum_{n=0}^{\infty} B_n \sin \frac{n\pi}{a} x e^{-\alpha \left(\frac{n\pi}{a}\right)^2 t}$$

$$T_1(x,0) = \sum_{n=0}^{\infty} B_n \sin \frac{n\pi}{a} x = f(x)$$

Multiplying both sides by $\sin m\pi/a\ x$ and integrating from 0 to a,

$$\int_0^a \sin \frac{m\pi}{a} x \sum_{n=0}^{\infty} B_n \sin \frac{n\pi}{a} x dx = \int_0^a f(x) \sin \frac{m\pi}{a} x dx$$

The left hand term may be represented by an orthogonal function as follows:

$$\int_0^a \sin mx \sin nx \, dx = 0; \; m \neq n; \; m,n = 0, \pm 1, \pm 2,\ldots$$

$$\int_0^a \sin mx \sin nx \, dx \neq 0; \; m = n; \; m,n = 0, \pm 1, \pm 2,\ldots$$

$$\int_0^a \sin mx \left[B_1 \sin \frac{\pi}{a} x + B_2 \sin \frac{2\pi}{a} x + \ldots + B_n \frac{n\pi}{a} x \right] dx = \int_0^a f(x) \sin \frac{m\pi}{a} x dx$$

when $m \neq n$, the left hand term equals 0; when $m = n$

$$\int_0^a B_n \sin \frac{n\pi}{a} x \sin \frac{n\pi}{a} x dx = \int_0^a f(x) \sin \frac{n\pi}{a} x dx$$

$$\int_0^a B_n \sin^2 \frac{n\pi}{a} x dx = \int_0^a f(x) \sin \frac{n\pi}{a} x dx$$

$$B_n \int_0^a \sin^2 \frac{n\pi}{a} x dx = \int_0^a f(x) \sin \frac{n\pi}{a} x dx$$

$$B_n = \frac{\displaystyle\int_0^a f(x) \sin \frac{n\pi}{a} x dx}{\displaystyle\int_0^a \sin^2 \frac{n\pi}{a} x dx}$$

$$T_1(x,t) = \sum_{n=0}^{\infty} \left[\frac{\displaystyle\int_0^a f(x) \sin \frac{n\pi}{a} x dx}{\displaystyle\int_0^a \sin^2 \frac{n\pi}{a} x dx} \right] \sin \frac{n\pi}{a} x e^{-\alpha \left(\frac{n\pi}{a}\right)^2 t}$$

$$\int_0^a \sin^2 \frac{n\pi}{a} x dx = \int_0^a dx - \int_0^a \cos^2 \frac{n\pi}{a} x dx$$

$$= \left[x \right]_0^a - \left[\frac{1}{2} x + \frac{1}{2\left(\frac{n\pi}{a}\right)} \sin \frac{n\pi}{a} x \cos \frac{n\pi}{a} x \right]_0^a$$

$$= a - \frac{a}{2} + \frac{2a}{n} \sin n\pi \cos n\pi$$

$$= \frac{a}{2} + \frac{2a}{n\pi} \sin n\pi \cos n\pi = \frac{a}{2}$$

$$T_1(x,t) = \sum_{n=0}^{\infty} \frac{\int_0^a f(x) \sin \frac{n\pi}{a} x dx}{a/2} \sin \frac{n\pi}{a} x e^{-\alpha\left(\frac{n\pi}{a}\right)^2 t}$$

$$T_1(x,t) = \frac{2}{a} \sum_{n=0}^{\infty} \left[\int_0^a f(x) \sin \frac{n\pi}{a} x dx \right] \sin \frac{n\pi}{a} x e^{-\alpha\left(\frac{n\pi}{a}\right)^2 t}$$

For $f(x) = 1$

$$\int_0^a \sin \frac{n\pi}{a} x dx = \left[-\frac{a}{n\pi} \cos \frac{n\pi}{a} x \right]_0^a$$

$$= -\frac{a}{n\pi} \cos n\pi + \frac{a}{n\pi} = \frac{a}{n\pi} (1 - \cos n\pi)$$

$$= \frac{2a}{n\pi} \text{ for n odd}$$

$$T_1(x,t) = \frac{2}{a} \sum_{n=0}^{\infty} \frac{2a}{n\pi} \sin \frac{n\pi}{a} x e^{-\alpha\left(\frac{n\pi}{a}\right)^2 t}$$

$$T_1(x,t) = \frac{4}{n\pi} \sum_{n=0}^{\infty} \sin \frac{n\pi}{a} x e^{-\alpha\left(\frac{n\pi}{a}\right)^2 t} \tag{2}$$

From this equation, we may deduce an equation for $T_1(x,y,z,t)$ for a rectangular parallelepiped of dimensions a,b,c

$$T_1(x,y,z,t) = \frac{4}{n\pi} \frac{4}{m\pi} \frac{4}{p\pi} \sum_{n=0}^{\infty} \sum_{m=0}^{\infty} \sum_{p=0}^{\infty} \sin \frac{n\pi}{a} x \sin \frac{m\pi}{b} y \sin \frac{p\pi}{c} z$$

$$e^{-\alpha\left(\frac{n^2\pi^2}{a^2} + \frac{m^2\pi^2}{b^2} + \frac{p^2\pi^3}{c^2}\right) t}$$

$$T_1(x,y,z,t) = \frac{64}{nmp\pi^3} \sum_{n=0}^{\infty} \sum_{m=0}^{\infty} \sum_{p=0}^{\infty} \sin \frac{n\pi}{a} x \sin \frac{m\pi}{b} y \sin \frac{p\pi}{c} z$$

$$e^{-\alpha\left(\frac{n^2\pi^2}{a^2} + \frac{m^2\pi^2}{b^2} + \frac{p^2\pi^2}{c^2}\right) t} \tag{3}$$

Where n, m, p are odd numbers for x, y, and z respectively in the intervals $0 < x < a$, $0 < y < b$, $0 < z < c$.

This equation is governed by the following initial and boundary conditions:

$$T_1(x,y,z,0) = 1 \tag{4}$$

$$T_1(a,y,z,t) = T_1(0,y,z,t) = 0 \tag{5}$$

$$T_1(x,b,z,t) = T_1(x,0,z,t) = 0 \tag{6}$$

$$T_1(x,y,c,t) = T_1(x,y,0,t) = 0 \tag{7}$$

Condition (4) establishes that the initial temperature is constant, and conditions (5), (6), and (7), that the temperature at the surface of the product is constant.

A change of variable of the following order is necessary in order to obtain the true temperature at any point of the container as a function of time, t:

$$T_1(x,y,z,t)_R = T_1(x,y,z,t)(T_o - T_h) + T_h \tag{8}$$

where $T_1(x,y,z,t)_R$ = true temperature at any point of the container as a function of time, t, during the heating period. $T_1(x,y,z,t)$ = temperature obtained from Equation 3; T_o = initial temperature of the product; T_h = temperature of the heating medium.

From Equation 8 we may calculate the true initial and boundary conditions:

$$T_1(x,y,z,t)_R = T_o \text{ when } T_1(x,y,z,0) = 1 \tag{9}$$

$$T_1(a,y,z,t)_R = T_1(0,y,z,t)_R = T_h \text{ when } T_1(a,y,z,t) = T_1(0,y,z,t) = 0 \tag{10}$$

$$T_1(x,b,z,t)_R = T_1(x,0,z,t)_R = T_h \text{ when } T_1(x,b,z,t) = T_1(x,0,z,t) = 0 \tag{11}$$

$$T_1(x,y,c,t)_R = T_1(x,y,0,t)_R = T_h \text{ when } T_1(x,y,c,t) = T_1(x,y,0,t) = 0 \tag{12}$$

C. Temperature Prediction During the Cooling Period

The temperature at any point in the rectangular pouch for a time, t, greater than the total heating time (t > fht) is obtained by solving the heat conduction equation as it was done for the prediction of the temperature during the heating period.

Prediction of $T_2(x)$ during the cooling period:

$$t = 0$$
$$T_2(x,0) = T_2(x,fht) = f(x)$$

$$fht = \text{final heating time}$$

$$T_2(x,fht) = \frac{4}{n\pi} \sum_{n=0}^{\infty} \sin \frac{n\pi}{a} x e^{-\alpha\left(\frac{n\pi}{a}\right)^2 fht}$$

We had calculated before that

$$T(x,t) = \frac{2}{a} \sum_{n=0}^{\infty} \left(\int_0^a f(x) \sin \frac{n\pi}{a} x dx \right) \sin \frac{n\pi}{a} x e^{-\alpha \left(\frac{n\pi}{a} \right)^2 t}$$

$$T_2(x,t) = \frac{2}{a} \sum_{n=0}^{\infty} \int_0^a \left(\frac{4}{n\pi} \sum_{n=0}^{\infty} \sin \frac{n\pi}{a} x e^{-\alpha \left(\frac{n\pi}{a} \right)^2 fht} \sin \frac{n\pi}{a} x dx \right) \sin \frac{n\pi}{a} x$$

$$e^{-\alpha \left(\frac{n\pi}{a} \right)^2 t}$$

Let's evaluate

$$\int_0^a \frac{4}{n\pi} \sum_{n=0}^{\infty} \sin \frac{n\pi}{a} x e^{-\alpha \left(\frac{n\pi}{a} \right)^2 fht} \sin \frac{n\pi}{a} x dx$$

$$\frac{4}{n\pi} \int_0^a \sum_{n=0}^{\infty} \sin \frac{n\pi}{a} x e^{-\alpha \left(\frac{n\pi}{a} \right)^2 fht} \sin \frac{n\pi}{a} x dx$$

Dividing and multiplying by

$$e^{-\alpha \left(\frac{n\pi}{a} \right)^2 fht} \frac{4}{n\pi} \int_0^a \sum_{n=0}^{\infty} \frac{\sin \frac{n\pi}{a} x e^{-\alpha \left(\frac{n\pi}{a} \right)^2 fht}}{e^{-\alpha \left(\frac{n\pi}{a} \right)^2 fht}} \sin \frac{n\pi}{a} x e^{-\alpha \left(\frac{n\pi}{a} \right)^2 fht} dx$$

Representing the numerator by an orthogonal function for m = n, we get

$$\frac{4}{n\pi} \int_0^a \sin^2 \frac{n\pi}{a} x e^{-\alpha \left(\frac{n\pi}{a} \right)^2 fht} dx =$$

$$\frac{4}{n\pi} e^{-\alpha \left(\frac{n\pi}{a} \right)^2 fht} \int_0^a \sin^2 \frac{n\pi}{a} x dx =$$

$$\frac{4}{n\pi} e^{-\alpha \left(\frac{n\pi}{a} \right)^2 fht} \frac{a}{2} = \frac{2a}{n\pi} e^{-\alpha \left(\frac{n\pi}{a} \right)^2 fht}$$

$$T_2(x,t) = \frac{2}{a} \sum_{n=0}^{\infty} \frac{2a}{n\pi} e^{-\alpha \left(\frac{n\pi}{a} \right)^2 fht} \sin \frac{n\pi}{a} x e^{-\alpha \left(\frac{n\pi}{a} \right)^2 t}$$

$$T_2(x,t) = \frac{4}{n\pi} \sum_{n=0}^{\infty} e^{-\alpha \left(\frac{n\pi}{a} \right)^2 fht} \sin \frac{n\pi}{a} x e^{-\alpha \left(\frac{n\pi}{a} \right)^2 t} \tag{13}$$

From this equation we may deduct an equation for the prediction of the temperature, $T(x,y,z,t)$ during the cooling time for a rectangular parallelepiped of dimensions a,b,c

$$T_2(x,y,z,t) = \frac{4}{n\pi} \frac{4}{m\pi} \frac{4}{p\pi} \sum_{n=0}^{\infty} \sum_{m=0}^{\infty} \sum_{p=0}^{\infty} e^{-\alpha\left(\frac{n^2\pi^2}{a^2} + \frac{m^2\pi^2}{b^2} + \frac{p^2\pi^2}{c^2}\right)fht}$$

$$\sin\frac{n\pi x}{a} \sin\frac{m\pi y}{b} \sin\frac{p\pi z}{c} e^{-\alpha\left(\frac{n^2\pi^2}{a^2} + \frac{m^2\pi^2}{b^2} + \frac{p^2\pi^2}{c^2}\right)t} =$$

$$T_2(x,y,z,t) = \frac{64}{nmp\pi^3} \sum_{n=0}^{\infty} \sum_{m=0}^{\infty} \sum_{p=0}^{\infty} e^{-\alpha\left(\frac{n^2\pi^2}{a^2} + \frac{m^2\pi^2}{b^2} + \frac{p^2\pi^2}{c^2}\right)fht} \sin\frac{n\pi x}{a}$$

$$\sin\frac{m\pi y}{b} \sin\frac{p\pi z}{c} e^{-\alpha\left(\frac{n^2\pi^2}{a^2} + \frac{m^2\pi^2}{b^2} + \frac{p^2\pi^2}{c^2}\right)t} \tag{14}$$

Where n,m,p are odd numbers for x,y, and z, respectively, the intervals $0 < x < a$, $0 < y < b$, $0 < z < c$.

This equation is governed by the following initial and boundary conditions:

$$T_2(x,y,z,0) = f(x,y,z) \tag{15}$$

$$T_2(0,y,z,t) = T_2(0,y,z,t) = 0 \tag{16}$$

$$T_2(x,b,z,t) = T_2(x,0,z,t) = 0 \tag{17}$$

$$T_2(x,y,c,t) = T_2(x,y,0,t) = 0 \tag{18}$$

Since the distribution of the temperature is not uniform at the end of the heating period, the initial distribution of the temperature during the cooling period is given by Equation 3, where $t = fht$, which is a function $f(x,y,z)$.

A change of variable of the following order is necessary in order to obtain the true temperature at any point of the container during the cooling period:

$$T_2(x,y,z,t)_R = T_2(x,y,z,t) + TCW \tag{19}$$

where $T_2(x,y,z,t)_R$ = true temperature at any point of the container as a function of time, t, during the cooling period. $T_2(x,y,z,t)$ = temperature obtained from Equation 14; TCW = average temperature of the cooling water.

D. Prediction of Nutrient Retention at any Point of the Container

In the case of nutrients, enzymes, colors, and microorganisms whose thermal destruction follow first order kinetics, the following reaction must hold:[10,39]

$$-\frac{dC(x,y,z,t)}{dt} = kC(x,y,z,t) \tag{20}$$

where k = reaction velocity constant which is a function of

$$T_i(x,y,z,t)$$

$C(x,y,z,t)$ = nutrient concentration at any point of the pouch at any time, t.
$T_i(x,y,z,t)$ = temperature at any point of the pouch at any time, t.

$$-\frac{dC(x,y,z,t)}{C(x,y,z,t)} = kdt$$

As it has been demonstrated by Daniels[10]

$$k = \frac{2.303}{D} = \frac{\ln 10}{D}$$

where D = time required to destroy 90% of the nutrient at a temperature, $T_i(x,y,z,t)$. $T_i(x,y,z,t)$ = temperature at any point of the container at any time, ϕ; ϕ = total processing time (heating and cooling).

According to Stumbo[43] and for a reaction velocity constant which follows the Arrhenius equation:

$$\log \frac{D}{D_o} = \frac{250 - T_i(x,y,z,t)}{Z_o} \tag{21}$$

$$D = D_o \, 10^{\left[\dfrac{250 - T_i(x,y,z,t)}{Z_o}\right]} \tag{22}$$

where D_o = time required to destroy 90% of the nutrient at a temperature of 250°F. Z_o = number of degrees Fahrenheit required for the thermal destruction curve to traverse one log cycle; mathematically equal to the reciprocal of the slope of the thermal destruction curve.

So,

$$k = \frac{\ln 10}{D_o \, 10^{\left[\dfrac{250 - T_i(x,y,z,t)}{Z_o}\right]}}$$

Substituting for k

$$-\frac{dC(x,y,z,t)}{C(x,y,z,t)} = \frac{\ln 10}{D_o \, 10^{\left[\dfrac{250 - T_i(x,y,z,t)}{Z_o}\right]}} dt$$

Which is the same as:

$$\frac{dC(x,y,z,t)}{C(x,y,z,t)} = -\frac{\ln 10}{D_o} e^{\left(\dfrac{T_i(x,y,z,t) - 250}{Z_o}\right)} \ln 10 \, dt \tag{23}$$

Integrating between C_o and $C(x,y,z,t)$, and between 0 and ϕ

$$\int_{C_o}^{C(x,y,z,t)} \frac{dC(x,y,z,t)}{C(x,y,z,t)} = -\frac{\ln 10}{D_o} \int_0^\phi e^{\left(\dfrac{T_i(x,y,z,t) - 250}{Z_o}\right)} \ln 10 \, dt$$

where C_o = initial concentration of the nutrient constant.

$$\frac{\ln C(x,y,z,t)}{C_o} = -\frac{\ln 10}{D_o} \int_0^\phi e^{\left(\frac{T_i(x,y,z,t) - 250}{Z_o}\right)} \ln 10 \, dt \frac{C(x,y,z,t)}{C_o}$$

$$= \exp\left[-\frac{\ln 10}{D_o} \int_0^\phi \exp\left(\frac{T_i(x,y,z,t) - 250}{Z_o}\right) \ln 10 \, dt\right] \tag{24}$$

where $C(x,y,z,t)/C_o$ = remaining fraction of nutrient at any point of the pouch of coordinates (x,y,z) at a time, ϕ.

This equation calculates the fraction of the nutrient remaining in one point of the pouch of coordinates (x,y,z) at the end of processing (heating and cooling).

E. Development of the Equation for the Prediction of Nutrient Retention After Thermal Processing

In order to calculate the quantity of nutrient retained at any point of the container at any time of processing, we plugged Equations 3 and 14 into Equation 24.

At the end of the heating period (t = fht), the concentration of the nutrient in each point of the container, $C(x,y,z,t)$ will vary due to the decreasing concentration gradient from the center to the walls of the container. The nutrient concentration at each point of the container at the end of the heating period will be equal to the initial concentration of the nutrient at the end of the cooling period.

In order to calculate $C(x,y,z,fht)$ we make ϕ = fht and substitute $T_1(x,y,z,t)$ for $T_i(x,y,z,t)$ in the previous equation. $C(x,y,z,fht)$ gives the concentration at the end of the heating period.

$$\frac{C(x,y,z,fht)}{C_o} = \exp\left[-\frac{\ln 10}{D_o} \int_0^{fht} \exp\left(\frac{T_i(x,y,z,t) - 250}{Z_o}\right) \ln 10 \, dt\right] \tag{25}$$

For the cooling period:

$$\frac{C(x,y,z,\phi - fht)}{C_o} = \exp\left[-\frac{\ln 10}{D_o} \int_0^{\phi - fht} \exp\left(\frac{T_2(x,y,z,t) - 250}{Z_o}\right) \ln 10 \, dt\right] \tag{26}$$

For the heating and cooling period:

$$\frac{C(x,y,z,t)}{C_o} = \exp\left[-\frac{\ln 10}{D_o} \int_0^{fht} \exp\left(\frac{T_1(x,y,z,t) - 250}{Z_o}\right) \ln 10 \, dt\right]$$
$$\exp\left[-\frac{\ln 10}{D_o} \int_0^{\phi - fht} \exp\left(\frac{T_2(x,y,z,t) - 250}{Z_o}\right) \ln 10 \, dt\right] \tag{27}$$

where ϕ − fht = total cooling time.

Equation 27 gives the fraction of nutrient which is not degraded at one point of the container of coordinates (x,y,z) at the end of processing.

Taking an average of the fraction of nutrient remaining in the pouch after a total processing time, ϕ:

$$\frac{\overline{C}}{C_o} = \frac{1}{abc} \int_0^a \int_0^b \int_0^c \frac{C(x,y,z,t)}{C_o} \, dx \, dy \, dz \tag{28}$$

$$\frac{\overline{C}}{C_o} = \frac{1}{abc} \int_0^a \int_0^b \int_0^c \exp\left[-\frac{\ln 10}{D_o} \int_0^{fht} \exp\left(\frac{T_1(x,y,z,t) - 250}{Z_o} \right) \ln 10 \, dt \right]$$

$$\exp\left[-\frac{\ln 10}{D_o} \int_0^{\phi - fht} \exp\left(\frac{T_2(x,y,z,t) - 250}{Z_o} \right) \ln 10 \, dt \right] dx \, dy \, dz \qquad (29)$$

Equation 29 represents the model for the prediction of nutrient retention in foods packed in retortable pouches of rectangular shape after being sterilized by thermal conduction. This equation may also be applicable for the prediction of the retention of some colors, enzymes, and other compounds which degrade in accordance with first order kinetics.

A computer program was developed in order to solve Equation 29. This program is shown in Appendix A.

IV. MODEL VALIDATION

After the development of the model, a set of experiments was conducted in order to verify its validity. The experiments were conducted on a simulated food in which thiamin was included as one of its components with the purpose of determining its destruction after thermal processing.

Thiamin was selected because it is thermolabile and its degradation follows first order kinetics. This degradation was studied by Feliciotti and Esselen[11] over the temperature range of 228°F to 300°F and over the pH range of 4.5 to 7.0. They established that the most pronounced changes occur at a pH between 6.0 and 6.5. A pH of 6.0 was selected throughout the experiments for this particular reason and because the kinetic parameters D_o and Z_o have been calculated by Barreiro et al.[6] who corrected the values obtained by Mulley et al.[34,35] The values obtained were: $D_o = 157.2$ min and $Z_o = 45°F$.

The simulated food containing the vitamin, was packaged in a retortable pouch and processed in a still retort. Subsequently, the quantity of vitamin before and after processing was determined in order to obtain the fraction of the nutrient retained after the sterilization period.

A. Simulated Food

The simulated food was prepared by mixing corn starch, carboxymethylcellulose and thiamin in a pH 6.0 buffer solution of potassium phosphate, monobasic, at 0.1 N and sodium hydroxide at 0.1 N. This mixture had a solid pasty appearance and was heated by conduction as the model requires.

The following procedure was accomplished for the formulation of the simulated food:

1.   The pH 6.0 buffer was prepared by mixing 5.6 mℓ of 0.1 *N* sodium hydroxide for each 50 mℓ of 0.1 *N* potassium phosphate, monobasic.
2.   The buffer solution was placed in a Waring® blender and the carboxymethylcellulose (supplied by Hercules, Inc., New York) at a proportion of 1.75%, was added very slowly while stirring constantly.
3.   The corn starch (USP grade) was dissolved in a small quantity of buffer and then added to the mixture. Enough starch was added in order to get a 6% proportion in the final mixture.
4.   After a homogenous mixture was obtained, the vitamin was added at a proportion of 10 mg per 100 g of mixture.
5.   The mixture was homogenized again for uniform distribution of the vitamin throughout, and immediately heated on a stove until it reached the boiling point (197.6°F) with the purpose of gelatinizing the starch.
6.   The mixture was packaged into the retortable pouches, sealed, and placed into a refrigerator where it was maintained until processing (about 24 hr).

The thermal diffusivity of this mixture was determined by Herrera[24] utilizing the formula developed by Olson and Jackson[36] for cylindrical containers. The value obtained was 0.01576 in.^{2}/min.

B. Retortable Pouches

The retortable pouches used to conduct the experiments were supplied by Continental Flexible Packaging, a division of Continental Can Company, U.S.A. According to the manufacturer, these pouches are produced from a lamination of polyester, aluminum foil, and a polyolefin blend. In this particular case, a lamination of 48 polyester, 70-gauge aluminum foil and 300 polypropylene was used.

The Pantry Pack®, as these pouches are called by the manufacturer, has been fully approved by the Food and Drug Administration and the U.S. Department of Agriculture. The current FDA approval limits the maximum temperature at which these pouches may be processed at 250°F, although they resist higher temperatures, i.e., 270°F.

The size of the pouches utilized in the experiments was 5 1/2" × 7" with a capacity of 9 oz and 4 3/4" × 7 1/4" with a capacity of 6 oz. The thickness of the pouches was measured in the center of the pouch after they were filled and sealed. In the case of the 5 1/2" × 7" pouch filled with 9 oz of the simulated food, the thickness at the geometrical center of the pouch was 0.819 in. and that of the 4 3/4" × 7 1/4" pouch was 0.709 in.

A correction was made to these dimensions in which the seals forming the edges of the pouch were subtracted from the above-mentioned dimensions. The new dimensions obtained were 4.791;" × 6.252" × 0.819;" for the 5 1/2" × 7" pouch and 3.976" × 6.437" × 0.709" for the 4 3/4" × 7 1/4" pouch. The corrected values were used for the computer solution of the model.

C. Thermal Processing

After the simulated food was packaged into the pouches and cooled overnight in a refrigerator, the samples were thermally processed in a vertical still retort ("Dixie", model RDSW-3). Besides the automatic pressure control, this model is capable of supplying an overriding air pressure to the chamber whenever an additional pressure is needed during processing in order to prevent rises in the pressure inside the container caused by temperature changes during sterilization.

The retort procedure used for the thermal processing of the pouches is described as follows:

1. The retort was loaded, closed, and the vents opened.
2. The steam was turned on.
3. When the retort reached the desired temperature the process timing started.
4. At the end of the process time, the vents were closed, air at a pressure of 17 psig was admitted into the retort, the steam valve was closed, and the cooling water was admitted from the bottom.
5. The pressure was released and the process terminated when the temperature in the center of the pouch decreased to 100°F.

The pouches were removed quickly from the retort and placed in a water and ice mixture to avoid further destruction of the vitamin, and then placed in a refrigerator over night prior to the vitamin analysis. Several of the pouched samples of simulated food were not processed, instead they were kept in refrigeration until the analysis of the vitamin. These samples served as control in order to determine the initial quantity of thiamin present in the food.

D. Heat Penetration Studies

The heating characteristics of a food product in a package are developed from time-temperature data measured by a thermocouple located at the point of the slowest heating which is usually the geometric center of the container.

Table 1
EXPERIMENTAL DESIGN

Experiment	Size (in.)	Processing temp. (°F)	Processing time (min.)
1	5 1/2 × 7	246.7	27.5
2	5 1/2 × 7	240.0	57.0
3	5 1/2 × 7	248.5	46.9
4	4 3/4 × 7 1/4	238.5	69.4

The heat penetration studies were performed following the procedure detailed by Altrand and Ecklund.[1] The package gland unit was installed at the midpoint on one of the sides of the pouch and a copper-constant thermocouple (Ellab, model TG 67) was threaded through the unit.

The time-temperature relationships at the center of the pouch, as well as the temperature and time of heating, the temperature and time of cooling, and the total processing time were recorded by means of a time-temperature recorder (Leeds and Northrup, Speedomax model).

E. Experimental Design

For the purpose of verifying the validity of the model, four different experiments were conducted in order to obtain the vitamin retained after processing. The four experiments were organized utilizing different combinations of time, temperatures, and size of pouches, as detailed in Table 1. The processing times and temperatures were chosen arbitrarily, according to the sterilization procedures that are ordinarily used for retortable pouches.

A total of five pouches were used for each experiment. One of the five pouches was saved in the refrigerator as control and was not processed. The remaining four pouches were processed according to the time-temperature combination of the particular experiment. One of the four processed pouches was used for the heat penetration study and discarded at the end of the processing period.

After processing, the thiamin content of the samples was determined to obtain the fraction of the vitamin retained after processing.

The analysis of the vitamin in the control sample was conducted on four replicates obtained after three aliquots were taken from the sample. A similar procedure was used for the processed samples in which four replicates were analyzed for each of the three samples. This made a total of 12 determinations for the control and 12 determinations for the processed samples.

F. Vitamin Analysis

The quantities of vitamin present in the simulated food before and after processing were determined by the method recommended by the Association of Official Analytical Chemists (1975).

V. MODEL SENSITIVITY

The mathematical model proposed in this investigation was evaluated experimentally in order to verify its validity.

The experiments were conducted by determining the fraction of thiamin retained in a simulated food, packaged in a retortable pouch, and processed thermally in a still retort.

During the thermal processing of the pouches, the heat penetration curves were recorded to obtain the processing variables necessary as inputs to the computer for solving the equation and also for the prediction of the fraction of vitamin retained in each particular run. The different processing variables used in the different experiments are detailed in Table 2. These particular temperatures and times of processing were selected according to the common procedures for the sterilization of retortable pouches.

Table 2
PROCESSING CHARACTERISTICS OF THE DIFFERENT EXPERIMENTS

Experiment	Initial temp. (°F)	Processing temp. (°F)	Temp. of cooling water (°F)	Heating time (min.)	Cooling time (min.)
1	60	246.0	90	16.5	11.0
2	60	240.0	90	39.0	18.0
3	60	248.5	85	28.3	18.6
4	78	238.5	90	60.4	9.0

Table 3
PREDICTED AND EXPERIMENTAL TEMPERATURES AT THE END OF
THE HEATING PERIOD

Experiment	Processing temp. (°F)	Heating time (min.)	Temp. in the center of the pouch (°F)		% Dev.
			Predicted	Experimental	
1	246.0	16.5	241.1	238.0	1.30
2	240.0	39.0	240.0	239.8	0.08
3	248.5	28.3	248.2	247.0	0.50
4	238.5	60.4	238.5	238.5	0.00

Table 4
PREDICTED AND EXPERIMENTAL TEMPERATURES AT THE END OF
THE COOLING PERIOD

Experiment	Processing temp. (°F)	Cooling period (min.)	Temperature in the center of the pouch (°F)		% Dev.
			Predicted	Experimental	
1	246.0	11.0	103.6	100.0	3.6
2	240.0	18.0	92.8	91.0	2.0
3	248.5	18.6	87.7	95.0	7.7
4	238.5	9.0	106.7	91.0	16.0

A. Temperature Prediction at the Center of the Pouch

It has been shown that the fraction of the nutrient retained is dependent on the processing temperature of the food.[34,35] Consequently, the model predicts the temperatures at any point of the container during the heating and cooling periods prior to the prediction of the fraction of nutrient retained after processing.

The heat penetration curves were obtained at the center of the pouch, since this is the point of slowest heating of the container and the temperature at this particular point is used for the calculation of the processing times during commercial sterilization of food products.

The experimental temperatures obtained at the end of the heating and cooling periods are compared to the model predictions in Tables 3 and 4. These tables show the percent deviation between the predicted and experimental temperatures at the end of the heating and cooling periods for each experiment.

The percent deviations of the predicted and experimental temperatures at the end of the heating period ranged from 0.0 to 1.3% which demonstrates that the model is reliable in making the predictions. The percent deviations at the end of the cooling period were higher, ranging from 2 to 16.0%.

Table 5
PREDICTED AND EXPERIMENTAL FRACTIONS OF THIAMIN RETAINED
AFTER THERMAL PROCESSING

Experiment	Size of the pouch (in.)	Processing temp. (°F)	Total processing time (min.)	Fraction of thiamin retained		90% Confidence intervals of the average experimental values
				Predicted	Experimental	
1	5 1/2 × 7	246.0	27.5	0.878	0.929	± 0.054
2	5 1/2 × 7	240.0	57.0	0.747	0.719	± 0.039
3	5 1/2 × 7	248.5	46.9	0.737	0.767	± 0.044
4	4 3/4 × 7 1/4	238.5	69.4	0.631	0.619	± 0.018

It may be observed from the data that the predicted values for the temperatures at the center of the container at the end of the heating and cooling periods are not significantly different than those obtained experimentally. The higher percent deviations at the end of the cooling period could be due to the assumption of a very high heat transfer coefficient at the surface of the pouch during the cooling period. The heat transfer coefficient, in this case, may be affected by the admission of air into the retort to raise the degree of pressurization. If too much air is admitted into the retort, the heat transfer to the contents may deteriorate. Also, if this pressure is released too fast, the air trapped inside the pouch expands, deforming it, and making it lose its rectangular parallelepiped form.

It may be concluded from these results that the values predicted by the model and those obtained experimentally compare favorably, thus showing the validity of the model for this type of prediction.

B. Prediction of Nutrient Retention

Table 5 presents a comparison between the predicted and experimental average fractions of thiamin retained after the thermal processing of the simulated food. This table also shows the intervals which contain the true mean fraction of vitamin retained at the end of processing with 90% confidence. These intervals were calculated by the statistical analysis of the results obtained from the experimental determination of the vitamin before and after thermal processing. The "t" distribution was used for the statistical evaluations.

The percent deviations of the average fraction of thiamin predicted by the model with respect to those obtained experimentally were 5.4% for the first experiment, 3.9% for the second, 3.9% for the third, and 1.9% for the fourth. These percent deviations show that the vitamin fractions predicted by the model are not significantly different than those obtained experimentally.

VI. CONCLUSIONS

The vitamin fractions predicted by the model were within the 90% confidence intervals obtained by statistical analysis of the experimental fractions from each determination. These results lead to the conclusion that the model is valid for the prediction of the retention of nutrients in foods packaged in retortable pouches and sterilized by thermal conduction which degrade according to first order kinetics.

This model may be used for the prediction of the degradation of colors, enzymes, and other compounds which degrade following the first order kinetics during thermal processing of foods in retortable pouches.

ACKNOWLEDGMENT

This chapter is based on the work performed by Mr. Monteza on his dissertation, "Prediction of Nutrient Retention in Foods Packaged in Retortable Pouches and Sterilized by Thermal Conduction".

Appendix A

```
         EXTERNAL FUNE
         COMMON /POI/ NPO,NPE
         COMMON /CONS/ PI, PI2,PI3,NH,NI,NP,PA,PB,PC,PA2,PB2,PC2
         COMMON A,B,C,XK,AA,XKO,TO,TV,TAP,TPC,TPE,TREF
C        MAIN PROGRAM
C        ******************************************************
C
         PI = 3.1415926
         IN = 5
         I00 = 6
C        ******************************************************
C
C        NUTRIENT LOSS DURING THERMAL PROCESSING OF FOODS PACKED
C        IN RETORTABLE POUCHES
C        TO = INITIAL TEMPERATURE OF THE PRODUCT (F)
C        TV = PROCESSING TEMPERATURE (F)
C        TAF = AVERAGE TEMPERATURE OF THE COOLING WATER (F)
C        VAL = VALUE OF THE TRIPLE INTEGRAL (C/CO)
C        T = PROCESSING TIME (MIN)
C        AA = INVERSE OF THE SLOPE OF THE THERMAL DEATH CURVE
C        XKO = DECIMAL REDUCTION TIME (TIME)
C        XK = THERMAL DIFFUSIVITY (IN**2/MIN)
C        NI = NO. OF POINTS FOR THE TRIPLE INTEGRAL
C        TFC = FINAL HEATING TIME (MIN)
C        TFE = FINAL COOLING TIME (MIN)
C        NPE = NO. OF POINTS FOR THE COOLING INTEGRAL
C        NPO = NO. OF POINTS FOR THE HEATING INTEGRAL
C        *************************************************************
C
         THEF = 25C.
         READ (IN,500)NPE,NPC,NI,NN,NM,NP
500      FORMAT(6I10)
         READ(IN,5 10)A,B,C
510      FORMAT(3F10.5)
         PA = PI/A
         PB = PI/B
         PC = PI/C
         PA2 = PA * PA
         PB2 = PB * PB
         PC2 = PC * PC
         PI2 = PI * PI
         PI3 = 64./(PI2 * PI)
         READ(IN,520)XK,XKC,AA
520      FORMAT(3F20.10)
         XKO = 2.302585/XKC
         AA = 0.4342944 * AA
         READ(IL,530)TO,TV,TAF,TFC,TFE
530      FORMAT(5F10.5)
         VAL = 1.
         TCEN = TEMC(A/2.,B/2.,C/2.,TFC)
         WRITE(I00,100)
106      FORMAT(1H,5A,'A',11X,'B',11X,'C')
         WRITE(I00,107) A,B,C
107      FORMAT(1H,1X,3(F11.5))
         WRITE(I00,101)
101      FORMAT(5X,'TO',15X,'TV',16X,'XK',15X,'VAL',11X,'TFC',112X,'TCEN')
         WRITE(I00,100) TO,TV,XK,VAL,TFC,TCEN
100      FORMAT(1H,1X,6(F11.5,5X))
         WRITE(I00,103)
103      FORMAT(1H,5A,'TAF',15X,'XK',14X,'XKO',12X,'AA',14X,'VAL'.114X,'TFE',12X,'TCEN')
         CALL INMUL(FUNE,A,B,C,VAL,NI)
         TCEN = TE(A/2.,B/2.,C/2.,TFE)
         AA = AA/0.4342944
         XKO = 2.302565/XKO
200      FORMAT(10(I5,5X))
         WRITE (I00,108) TAF,XK,XKO,AA,VAL,TFE,TCEN
108      FORMAT(1H,1X,7(F11.5,5X))
         WRITE(I00,200)NPC,NPE,NI
         STOP
         END
```

```
       SUBROUTINE INMUL(FUNE,A,B,C,VAL,NI)
       DIMENSION R(30)
       SUM = 0.
       ISEED = 123457
       N = 30
       CALL GGOB (ISEED,N,R)
       DO 1 I = 1,NI
       U = A * R(I)
       V = B * R(I + 10)
       W = B * R(I + 20)
       SUM = SUM + FUNE(U,V,W)
1      CONTINUE
       VAL = SUM/FLOAT(NI)
       RETURN
       END
       FUNCTION TE(U,V,W,TIM)
C      TE = TEMPERATURE DURING COOLING (F)
       COMMON /CONS/ PI,PI2,PI3,NN,NM,NP,PA,PB,PC,PA2,PB2,PC2
       COMMON A,B,C,XK,AA,XKO,TO,TV,TAF,TPE,TREF
       COMMON/POI/NPC,NPE
       INTEGER P
       SUM = 0.
       DO 1 N = 1,NN,2
       N2 = N * N
       DO 1 M =1,NM,2
       M2 = M * M
       DO = P=1,NP,2
       P2 = P * P
       ZZZ = XK * PA2 + M2 * PB2 + P2 * PC2)
       AC = (EXP(-ZZZ * TFC) * (TO-TV) + (TV-TAF)/FLOAT (N * M * P)
       SUM = SUM + AC * SIN(N * PA * U) * SIN(M * PB * V)
       1* SIN(P PC * W) * EXP(-ZZZ * TIM)
1      CONTINUE
       TE = PI3 * SUM + TAF
       RETURN
       END
       FUNCTION FUNE(U,V,W)
       DIMENSION X(20),Y(20)
       COMMON/POI/ NPC,NPE
       COMMON A,B,C,XK,AA,XKO,TO,TV,TAF,TFC,TFE,TREF
       DELT = 1./FLOAT(NPE-1) * TFE
C      FUNE CALCULATES THE GENERAL CONCENTRATION FOR THE COOLING TIME X(1) = 0.
C      THE INTEGRAL'S INFERIOR LIMIT ABOUT THE TIME = 0
       Y(1) = EXP((TEMO)U,V,W,TFC) - TREP)/AA)
C      THE TIME CORRESPONDS TO THE FINAL HEATING TIME
       DO 1 I=2,NPE
       X(I) = X(I-1) + DELT
       Y(I) = EXP((TE(U,V,W,X(I)) - TREF)/AA
1      CONTINUE
       FF = AVIN1(X,Y,NPE,X(I),X(NPE))
       FUNE = EXP(-XKO * FF) * FUNC(U,V,W,TFC)
       RETURN
       END
       FUNCTION TEMO(U,V,W,TIM)
       COMMON /POI/ NPC,NPE
       COMMON A,B,C,XK,AA,XKO,TO,TV,TAF,TFC,TFE,TREF
       COMMON /CONS/ PI,PI2,PI3,NW,NM,NP,PA,PB,PC,PA2,PB2,PC2
       INTEGER P
       SUM = 0
       DO 1 B=1,NM,2
       M2 = M * M
       DO 1 N=1,NN,2
       N2 = N * N
       DO 1 P=1,NP,2
       P2 = P * P
       Q = EXP(-XK * (N2 + PA2 + M2 * PB2 P2 * PC2) * TIM)
       SUM = SUM + 1./FLOAT(N * M * P) * SIN(N*PA*U) * SIN(M*PB*V) 1* SIN(P * PC W) * C
C      ************************************************************
C
C      CALCULATES THE TEMPERATURE DURING HEATING TIME (F)
C
C      ************************************************************
1      CONTINUE
```

```
              TEMC = (TO − TV) = PI3 * SUM + TV
              RETURN
              END
              FUNCTION FUNC (U,V,W,TIM)
              DIMENSION X(20),Y(20)
              COMMON /POI/ NPC,NFE
              COMMON A,B,C,XK,AA,XKO,TO,TV,TAF,TFC,TFE,TREF
              DELT = 1./FLOAT(NPC − 1) * TIM
C     ************************************************************
C     FUNC CORRESPONDS TO THE CONCENTRATION DURING HEATING
C     INTEGRATES USING AVINT (FF)
C     NPC CORRESPONDS TO THE NO. OF POINTS TO CALCULATE AVINT
C     ************************************************************
              X(1) = 0.
              Y(1) = EXP((TO − TREF)/AA)
C     TO = INITIAL TEMPERATURE
              DO L I=2,NPC
              X(I) = X(I−1) + DELT
C     X STORES THE INTERVALS AND Y THE ORDINATES
              Y(I) = EXP(TEMC(U,V,W,X(I)) − TREF)/AA)
1             CONTINUE
              FF = AVINT(X,Y,NPC,X(1),X(NPC))
              FUNC = EXP(− XKO * FF)
              RETURN
              END
              FUNCTION AVINT (X,Y,N,XLO,XUP)
              DIMENSION X(20), Y(20)
              AVINT = 0.
              IF (XLO.EQ.XUP) RETURN
C     INTEGRATION OF EXPERIMENTAL DATA
C     P. J. DAVIS RABINOVITZ
C     THE DATA IS PRESENTED IN ONE VECTOR
C     THE X VALUES IN INCREASING ORDER
              SUM = 0.
              SYL = XLO
              J = N
              IB = 2
              DO 1 I=1,N
              IF(X(I) − XLO) 1,17,17
1             IB = IB + 1
17            DO 2 I=1,N
              IF (XOP − X(J)) 2,18,18
2             J = J−1
18            J = J−1
              DO 3 JM = IB,J
              X1 = X(JM−1)
              X2 = X(JM)
              X3 = X(JM+1)
              TERM1 = Y(JM−1)/(X1−X2) * (X1−X3)
              TERM2 = Y(JM)/((X2−X1) * (X2−X3))
              TERM3 = Y(JM+1)/((X3−X1) * (X3−X2))
              A = TERM1 + TERM2 + TERM3
              B = −(X2 + X3) * TERM1 − (X1+X3) * TERM2 − (X1+X2) * TERM3
              C = X2 * X3 * TERM1 + X1 * TERM2 + X1 * X2 * TERM3
              IF (JM−IB) 14,4,14
4             CA = A
              CB = B
              CC = C
              GO TO 15
14            CA = 0.5 * (A + CA)
              CB = 0.5 * (B + CB)
              CC = 0.5 * (C + CC)
15            SYU = X(JM)
              SUM = SUM + CA * (SYU**3 − SYL**3)3/3. + CB * 0.5* (SYU**2 − SYI**
              12) + CC * (SYU − SYL)
              CA = A
              CB = B
              CC = C
3             SYL = SYO
              AVINT = SUM + CA * (XUP**3 − SYL**3)3. + CB * 0.5 * (XUP**2 − SYI 1**2) + CC * (XUP − SYL)
              RETURN
              END
```

Appendix B
LIST OF SYMBOLS

A: Constant

a: Dimensions of the container on the x axis

B: Constant

b: Dimension of the container on the y axis

C: Nutrient concentration at a point of the container (mg/g)

C_o: Initial concentration of the nutrient in the container, constant (mg/g)

$\overline{C}$: Concentration of the nutrient in the container (mg/g)

c: Dimension of the container on the z axis

D: Time required at any temperature to destroy 90% of the nutrient at T_i (min)

D_o: Time required at any temperature to destroy 90% of the nutrient at 250°F (min)

E_a: Activation energy (Btu/mol)

F_o: The equivalent of all heat considered with respect to its capacity to destroy nutrients at 250°F (min)

F_s: Integrated lethal or degradative capacity of heat received by all points in a container during process (min at 250°F)

f_h: Time required for the straight-line portion of the semilog heating or cooling curve to traverse 1 log cycle (min)

fct: Final cooling time (min)

fht: Final heating time (min)

g: Difference between retort temperature and the maximum temperature reached by the food at the point of concern (°F)

j: Lag factor

k: Reaction velocity constant (min^{-1})

L: Constant

M: Constant

m: Positive integer number

n: Positive integer number

p: Positive integer number

R: Subscript which denotes real conditions

Rg: Gas constant (Btu/mol °R)

T: Temperature (°F)

T_a: Absolute temperature (°R)

Tl_h: Temperature of the heating medium (°F)

T_i: Temperature at any point of the container (°F)

T_o: Initial temperature of the mixture, constant (°F)

T_1: Temperature at any point of the container during the heating period (°F)

T_2: Temperature at any point of the container during the cooling period (°F)

TCW: Average temperature of the cooling water (°F)

t: Time (min)

U: The equivalent of the lethal heat received by some designated point in the container during process (min at retort temperature)

x: Rectangular cartesian coordinate

y: Rectangular cartesian coordinate

Z_o: Number of Fahrenheit degrees required for the thermal destruction curve to traverse 1 log cycle (°F)

z: Rectangular cartesian coordinate

Greek Letters

α: Thermal diffusivity (in²/min)

φ; Total processing time period (min)

λ: Nondimensional parameter

θ: Time function

REFERENCES

1. **Altrand, D. V. and Ecklund, O. F.,** The mechanics and interpretation of heat penetration test in canned foods, *Food Technol.*, 6(5), 185, 1952.
2. **Anon.,** "Flexible can" scores for pre-cooked vegetables, *Modern Packaging*, 42(12), 82, 1969.
3. **Anon.,** Retort pouch nears market stage, *Food Eng.*, 44(5), 160, 1972.
4. Assoc. of Offic. Agric. Chem., *Official Methods of Analysis of the AOAC*, 12th ed., AOAC, Washington, D.C., 1975, 771.
5. **Ball, C. O. and Olson, F. C. W.,** *Sterilization in Food Technology*, McGraw Hill, New York, 1957.
6. **Barreiro, J. A., Salas, G. R., and Herrera, I.,** Formulacion y evaluacion de un modelo matematico para la prediccion de perdidas de nutrientes durante el procesamiento termico de alimentos enlatados, *Arch. Latinoam. Nutr.*, 27(3), 325, 1977.
7. **Barreiro, J. A., Salas, G. R., and Herrera, I.,** Optimizacion nutricional del procesamiento termico de alimentos enlatados, *Arch. Latinoam. Nutr.*
8. **Carslaw, H. S. and Jaeger, J. C.,** *Conduction of Heat in Solids*, 2nd ed., Oxford University Press, London, 1959.
9. **Charm, S. E.,** *Fundamentals of Food Engineering*, 3rd ed., AVI Publ., Westport, Conn., 1978.
10. **Daniels, F. and Alberty, R. A.,** *Physical Chemistry*, 4th ed., John Wiley & Sons, New York, 1970.
11. **Feliciotti, E. A. and Esselen, W. B.,** Thermal destruction rates of thiamine in pureed meats and vegetables, *Food Technol.*, 11(12), 77, 1957.
12. **Gillespy, T. G.,** Estimation of sterilizing value of process as applied to canned foods. II. Packs heating by conduction, complex processing conditions and value of coming-up time of retort, *J. Sci. Food Agric.*, 4, 553, 1953.
13. **Gould, W. A., Geisman, J. R., Weiser, H. H., Bash, W. D., Moore, W. H., Salzer, R. H., and Long, F. E.,** Heat processing of vegetables in flexible films. Research Bull. 905, Ohio Agricultural Experiment Station, Wooster, Ohio, 1962.
14. **Hayakawa, K. and Ball, C. O.,** A note on theoretical cooling curve of a cylindrical can of thermally conductive food, *Can. Inst. Food Technol. J.*, 2(3), 115, 1969.
15. **Hayakawa, K. and Ball, C. O.,** Charts for calculating average temperature of thermally conductive food in a cylindrical can during heat processing, *J. Inst. Technol. Aliment.*, 2(1), 12, 1969.
16. **Hayakawa, K. and Timbers, G. E.,** Transient state heat transfer in stacks of heat processed food stored in a commercial warehouse, *J. Food Sci.*, 41, 833, 1976.
17. **Hayakawa, K.,** Estimating food temperatures during various processing or handling treatments, *J. Food Sci.*, 36, 378, 1971.
18. **Hayakawa, K.,** Estimating temperatures of foods during various heating or cooling treatments, *Ashrae J.*, 9, 65, 1972.
19. **Hayakawa, K.,** Estimating the central temperature of canned food during the initial heating or cooling period of heat process, *Food Technol.*, 23(11), 141, 1969.
20. **Hayakawa, K.,** Experimental formulas for accurate estimation of transient temperature of food and their application to thermal process evaluation, *Food Technol.*, 24(12), 89, 1970.
21. **Hayakawa, K.,** New parameters for calculating mass average sterilizing values to estimate nutrients in thermally conductive foods, *Can. Inst. Food Technol. J.*, 2(4), 167, 1969.
22. **Hayakawa, K.,** Response for estimating temperatures in cylindrical cans of solid food subjected to time variable processing temperatures, *J. Food Sci.*, 39, 1090, 1974.
23. **Heldman, D. R.,** *Food Process Engineering*, reprint ed., AVI Publ., Westport, Conn., 1977.

24. **Herrera, I.,** Estudio Experimental de un Modelo Matematico para la Prediccion de Perdidas de Nutrientes durante el Procesamiento Termico de Alimentos Enlatados, Universidad Simon Bolivar, Caracas, Venezuela, 1975.

25. **Herrera, I.,** Prediccion de la Retencion de Nutrientes en los alimentos Calentados por Conduccion en Latas Rectangulares, Master's thesis, Universidad Simon Bolivar, Caracas, Venezuela, 1978.

26. **Jen, Y., Manson, J. E., Stumbo, C. R., and Zahradnik, J. W.,** A procedure for estimating sterilization of and quality factor degradation in thermally processed foods, *J. Food Sci.,* 36, 692, 1971.

27. **Jokinen, J. E. and Reineccius, G. A.,** Losses in available lysine during thermal processing of soy protein model systems, *J. Food Sci.,* 41, 816, 1976.

28. **Komatsu, Y.,** Technical information on flexible packaging of heat-processed food products, *New Food Ind.,* 13, 7, 1974.

29. **Kreith, F.,** *Principles of Heat Transfer,* 3rd ed., Intex Ed. Publ., New York, 1973.

30. **Laing, B. M., Schlueter, D. L., and Labuza, T. P.,** Degradation kinetics of ascorbic acid at high temperature and water activity, *J. Food Sci.,* 43(5), 1440, 1978.

31. **Lenz, M. K. and Lund, D. B.,** The lethality-Fourier number method: confidence intervals for calculated lethality and mass-average retention of conduction-heating, canned foods, *J. Food Sci.,* 42(12), 1002, 1977.

32. **Lenz, M. K. and Lund, D. H.,** The lethality-Fourier number method: experimental verification of a model for calculating average quality factor retention in conduction-heating canned foods, *J. Food Sci.,* 42(4), 997, 1977.

33. **Lenz, M. K. and Lund, D. B.,** The lethality-Fourier number method: experimental verification of a model for calculating temperature profiles and lethality in conduction-heating canned foods, *J. Food Sci.,* 42(4), 989, 1977.

34. **Mulley, E. A., Stumbo, C. R., and Hunting, W. H.,** Kinetics of thiamine degradation by heat. Effect of pH and form of vitamin on its rate of destruction, *J. Food Sci.,* 40, 989, 1975.

35. **Mulley, E. A., Stumbo, C. R., and Hunting, W. M.,** Thiamine: a chemical index of the sterilization efficacy of thermal processing, *J. Food Sci.,* 40, 993, 1975.

36. **Olson, F. C. W. and Jackson, J. M.,** Heating curves: theory and practical application, *Ind. Eng. Chem.,* 34, 337, 1942.

37. **Rees, J. A. G.,** Processing heat-sterilizable flexible packs, *FMF Rev.,* 9(1), 10, 1974.

38. **Saguy, I., Mizrahi, S., Villota, R., and Karel, M.,** Accelerated method for determining the kinetic model of ascorbic acid loss during dehydration, *J. Food Sci.,* 43(5), 1861, 1978.

39. **Stevens, B.,** *Chemical Kinetics,* Chapman and Hall, London, 1971.

40. **Stumbo, C. R.,** Bacteriological considerations relating to process evaluation, *Food Technol.,* 2, 116, 1948.

41. **Stumbo, C. R.,** Further considerations relating to evaluation of thermal process for foods, *Food Technol.,* 3, 126, 1949.

42. **Stumbo, C. R.,** New procedures for evaluating thermal procedures for evaluating thermal process for foods in cylindrical containers, *Food Technol.,* 7, 309, 1953.

43. **Stumbo, C. R.,** *Thermobacteriology in Food Processing,* Academic Press, New York, 1973.

44. **Teixeira, A. A., Dixon, J. R., Zahradnik, J. W., and Zinsmeister, G. E.,** Computer optimization of nutrient retention in the thermal processing of conduction heated foods, *Food Technol.,* 23(6), 845, 1969.

45. **Teixeira, A. A., Stumbo, C. R., and Zahradnik, J. W.,** Experimental evaluation of mathematical and computer models for thermal process evaluation, *J. Food Sci.,* 60, 652, 1975.

46. **Thorne, S.,** Retortable pouches, *Nutr. Food Sci.,* 45, 2, 1976.

47. **Wanninger, L. A.,** Mathematical model predicts stability of ascorbic acid in food products, *Food Technol.,* 26(6), 42, 1972.

48. **Wolf, J. C. and Thompson, D. R.,** Initial losses of available lysine in model systems, *J. Food Sci.,* 42(6), 1540, 1977.

49. **Woollen, A.,** What's new in Europe, *Food Eng.,* 44(33), 86, 1972.

50. **Yamaguchi, K., Komatsu, Y., and Kishimoto, A.,** Nisshokuhsi. Nippon Koyo Gakkai shi, *J. Food Sci. Technol.,* 18(2), 75, 1971.

INDEX

A

Absorption, 85
Activity, 3, 5
Albumin kinetics, 99
Alcohol, vapor toxicity of, 85
Aldosterone, 92
Algorithm, 16—19, 81, 89
  development, 14
Algorithmic level, 17, 22
Alternative models, 8
Analog computers, 2, 79, 88—89
Analog-model, 10
Anemia, 11
Animal limb movements, 10
Angiotensin, 92
Anti-arthritic activity, 82, 84
Antibiotics, 2
Antihypertensive drugs, 92
Aplastic anemia, 59, 62
Arrhenius equation, 115, 122
Arterial blood pressure
  distribution, 92
  elevation, 104
  Gaussian type distribution, 92
  mean value, 100
Artificial cell, 61, 65
Assumptions, 4, 15
Asynchronous state-controlled sequential circuit, 64
Attributes, 3—4, 14
Automata theory, 64
Autoregulation, 95

B

Bacterial growth, 2—3
Bacterial inactivation, 114
Barbiturates, 81
Baro-chemo system to stabilize blood pressure,
  96—97, 104
Baro-receptors, 96, 105
Basal response, 10
Bayes' theorem, see also Bayesian methods, 47—48
Bayesian methods, 36—38, 41, 44—45, 50
Behavioral models, 4
Bernoulli trial, 20, 22—24
Best Linear Unbiased Estimators (BLUE), 33
Binomial process, 20
Binomial random variable, 20
Biochemical reaction mechanisms, 78
Biocybernetic systems, 10
Biodynamic compartmental models, solution of, 89
Biodynamic system, 86
Biological activity, 78—81
Biological age of rat, 10
Biological oxygen demand (BOD), 30
Biological profile, 78

Biological response, 78
Biological system, 2—3
Biomedical effectors, computers in modeling, structure and function of, 10
Block-oriented languages, 89
Blood flow, 92, 94
Blood-forming process, 74
  cell renewal system, 57
  description of the model, 56, 58
  deterministic model, 59
  motivation for study, 56—58
  performance of simulation, 58—61
  regulatory model of, 56—62
  results of simulation, 59 62
Blood pressure, 92, 94
  arterial, see Arterial blood pressure
  baro-chemo system to stabilize, 96—97, 104
  epidemiological studies, 92
Blood volume, 92, 94, 104
  mean value, 100—101, 103
Body temperature, computers in determining effect
  on thermal regulation, 10—11
Boundaries, 3—5
Boundary condition, 116
Box and Draper's determinant criterion, 44
Box-Hill criterion, 45, 47—49
Branching proceses, theory of, 56
Break point, 20, 23
Bronchial dilator, 88

C

Cancer, 55—75
  first approach to modeling, 56—57
  multi-loop compartment model, 59, 61—63, 66—
    67
  regulatory model of blood-forming process, 56—
    62
  self reproducing normal and malignant cells, 63—
    65
  spread of cells in tissue, 65—74
Canned foods, 113
Carboxymethylcellulose, 124
Carcinogenesis, multi-stage hypothesis, 74
Carcinogenic chemicals, 56
Cardiac output, 11, 92, 94, 100—101, 103, 105
Cardiac output-blood volume, 100—101
Cardiac output-heart rate, 100
Cardiovascular model, 15
Cardivascular morbidity, 92
Cardiovascular research, see also specific topics
  computer modeling in, 91—110
  Guyton-Coleman model, 94—99
  hypertension, 92—94
Catalyst fouling system, 47
Catecholamines, 9, 99
Cell analysis in hematology, computers for, 11

Immunoregulatory activity, 82, 84
Impairment, 93
Implementation effort, 14
Independent variables, 4—5
Infusion, 93
Initial estimates, 35, 47
Input, 3, 23
Input channels, 61
Interacting subsystems, 58
Intrinsic parameters, 36
Ionizing radiation, 56
Iterative procedure, 34—35, 43

K

Kidney, 95, 105
  role in hypertension, 92
Kier index, 79, 83—86
Kinetic model building, 39
Kinetics of microorganism inactivation, 113
Kronecker product, 43
Kubinyi model, 79—81

L

Lack-of-fit F-tests, 39
Lactic acid, 8
Least squares estimates of parameters, 33
Least-squares method, 9
Lethality-Fourier number, 112, 114
Leukocyte Automatic Recognition Computer
  (LARC), 11
Lewis lung carcinoma, 81
Life span matrix element, 64, 68
Life span of cell, 68, 70
Likelihood discrimination, 38—39, 44—45, 50
Likelihood ratio method, 38—39
Linearized equations of Guyton-Coleman model, 97
Linear model, 80
  parameter estimation in, 33—34
Linear regression, 35
  confidence intervals, 34
  confidence regions, 34
  parameter estimation, 33—34
Linear transformations, 9
Lipophilicity, 78—79
Logical data structure language, 14
Loop structure, 23
Loss of nutrients, 112
  prediction of, 114
Low pressure zone, 95

M

Malignant cell growth, see also Spread of cancer
  cells in tissue, modeling of, 66, 74
Mass action law, 9
Mass average sterilizing value, 113

Mathematical description, 2, 14
Mathematical formulation of multiresponse models,
  42
Mathematical model, 2, 30
  biological age of rat, 10
  cell growth, 56
Mathematical modeling, 4—5
Mathematics, 16—19
Maximum likelihood, 45
Maximum likelihood function, 38
Maximum likelihood ratio, 38
Maxwell density function, 21
Mean life span, 61, 64
Mean substructure activity frequency (MSAF), 82,
  84
Mean systemic pressure, 95
Mean value
  arterial pressure, 100
  blood volume, 100—101, 103
  heart rate, 100—101, 103
  study of, 104—106
Mechanical systems, 2
Mechanical ventilation, 11
Mechanistic modeling, 31—33
Messenger ribonucleic acid, 19
Metabolic pathway, 7—9
Metabolic problems, 7
Metabolism, 85
Michaelis constant, 8
Microcomputers, role of, 89
Microorganisms, growth of, 2
Mild hypertension, 105
Mitotic inhibitors, 58
Model behavior, 14—15
Model building methods, 5—6
Model discrimination, see Discrimination
Model sensitivity in food processing, 126—128
Model validation in food processing, 124—126
Modeling, see also specific topics, 1—12, 14, 50
  benefits, 4
  cancer problems and its progress, 55—75
  cardiovascular research, 91—110
  computers in, 10
  criteria of biosystem, 14
  defined, 4
  drug design, 77—90
  food process engineering systems, 111—134
  goals of, 31
  software concepts, 14
  spread of cancer cells in tissue, 65—74
Modeling strategy, 31—33
Models, see also specific types
  descriptive, 4
  main function, 4
  performance of, 4
  physical, 4—5
  prediction of performance criteria of system, 4
  selfreproducing normal and malignant cells, 63—
  65
  tests of adequacy, 39
  types, 4—5